Cultural Evolution and Social Ontology

This volume explores connections between two growing and complementary fields of research: cultural evolutionary theory and social ontology. It sheds light on the ontological aspects of cultural evolution that so far have been largely neglected and raises questions for social ontology regarding the relevance of evolutionary aspects of social phenomena.

This volume shows that cultural evolutionary theory and social ontology are complementary disciplines that, while having their own subject matter and their own research questions, illuminate each other in interesting ways. While the contributors vary in their approach to the relationship between social ontology and cultural evolution, they explore their many common themes, including power, language, agency, interaction, and social institutions and roles. The chapters are divided into thematic sections organized around metaperspectives on cultural evolution and social ontology, the power concept in cultural evolution and social ontology, and themes of interactions and collectives in cultural and social evolution. Together, the chapters demonstrate how social ontology can provide critical tests of central assumptions in cultural evolutionary theory, and how cultural evolution can provide accounts of the origins of social entities.

Cultural Evolution and Social Ontology will appeal to researchers and graduate students working in cultural evolution, social ontology, metaphysics, social philosophy, philosophy of the social sciences, philosophy of biology, and philosophy of science, as well as those working in disciplines outside philosophy, such as anthropology and sociology.

Martina Valković is a visiting researcher at the Institute of Philosophy, Leibniz University Hannover, Germany, and at Radboud University Nijmegen, The Netherlands.

Thomas A.C. Reydon is Professor of Philosophy of Science and Technology in the Institute of Philosophy and the Centre for Ethics and Law in the Life Sciences (CELLS) at Leibniz University Hannover, Germany, and Associated Faculty in the Socially Engaged Philosophy of Science (SEPOS) group at Michigan State University, USA.

Routledge Studies in Contemporary Philosophy

Digital Development
Technology, Ethics and Governance
Edited by Mirko Farina, Xiao Yu and Jin Chen

Exploring Suicidal Ambivalence
Philosophical and Psychological Perspectives
Edited by René Baston and Martin Weichold

Reasoning in the Wild
The Public Works of Reason
Mariam Thalos

Normativity, Moral Reasoning, and Human Rights
Engaging with Contemporary Chinese Moral and Political Theory
Philippe Brunozzi

From Human Reasoning to Belief
An Empirical Account
Joshua Mugg

Against Qualia, For Behaviorism
J.M. Fritzman

Cultural Evolution and Social Ontology
Interdisciplinary Perspectives
Edited by Martina Valković and Thomas A.C. Reydon

For more information about this series, please visit: www.routledge.com/
Routledge-Studies-in-Contemporary-Philosophy/book-series/SE0720

Cultural Evolution and Social Ontology

Interdisciplinary Perspectives

Edited by Martina Valković and
Thomas A.C. Reydon

Routledge
Taylor & Francis Group
NEW YORK AND LONDON

First published 2026
by Routledge
605 Third Avenue, New York, NY 10158

and by Routledge
4 Park Square, Milton Park, Abingdon, Oxon, OX14 4RN

Routledge is an imprint of the Taylor & Francis Group, an informa business

For Product Safety Concerns and Information please contact our EU representative GPSR@taylorandfrancis.com. Taylor & Francis Verlag GmbH, Kaufingerstraße 24, 80331 München, Germany.

ISBN: 978-1-041-08121-0 (hbk)
ISBN: 978-1-041-08120-3 (pbk)
ISBN: 978-1-003-64384-5 (ebk)

DOI: 10.4324/9781003643845

Typeset in Sabon
by Apex CoVantage, LLC

Contents

Contributors

Karim Baraghith is a postdoctoral researcher at the Institute of Philosophy, Leibniz University Hannover. He is the principal investigator of the DFG-funded project "Graphs and Networks as Explanatory Tools in Cultural Evolutionary Theory". He obtained a PhD in philosophy and has held research posts at Heinrich Heine University Düsseldorf and Leibniz University Hannover. His research concentrates on cultural and generalized evolutionary theory, with a specific emphasis on network- and graph-based explanations, the integration of micro–macro models, population concepts, and the evolution of languages.

Åsa Burman is Reader (*Docent*) in Practical Philosophy at Stockholm University, Sweden, and an affiliated researcher at the Institute for Futures Studies. She is President of the International Social Ontology Society. Burman's newest book, *Nonideal Social Ontology: The Power View* (Oxford University Press, 2023), is about social phenomena such as class, structures, and social power, as well as the key differences between ideal and non-ideal social ontology.

Azita Chellappoo is Lecturer in Philosophy at The Open University, UK. Her research is broadly concerned with the intersection between the biological and the social. This includes epistemic and normative dimensions of evolutionary approaches to cultural phenomena, the role of social and political values in shaping scientific understanding of the social world, and the characterization and use of social categories such as race, gender, and fatness in the life sciences.

Bart Geurts is Professor of the Philosophy of Language and Logic at the University of Nijmegen. He has authored and co-authored publications on a wide variety of topics in semantics and pragmatics, including monographs on presupposition and conversational implicatures. Although his core business is semantic and pragmatic theory, he has been involved in experimental research as well and has contributed to various other fields, including the psychology of language, social cognition, and, in recent years, evolutionary pragmatics.

Frank Hindriks is Professor of Ethics, Social and Political Philosophy at the Faculty of Philosophy of the University of Groningen. He is a founding member of the International Social Ontology Society and one of the founding editors of the *Journal of Social Ontology*. He is a member of the Center of Philosophy, Politics, and Economics in Groningen and a fellow of an interdisciplinary research project on sustainable cooperation (SCOOP). His research concerns social ontology and collective ethics. He is the author of *The Structure of the Open Society* (Oxford University Press, 2025).

Finn Jordan is currently pursuing a degree in psychology at the University of Hildesheim, where he also works as a research assistant in the Working Group for Social, Organizational, and Business Psychology. He previously earned a bachelor's degree in philosophy and sociology at Heinrich Heine University Düsseldorf and worked as a research assistant in the Department of Theoretical Philosophy.

Richard Lauer is Visiting Assistant Professor of Philosophy at St. Lawrence University in Canton, New York. He received his PhD in philosophy from the University of Missouri, Columbia, in 2016, and specializes in the philosophy of science, in particular in the philosophy of social science. He has published on questions about the methodology of social ontology (and its relation to social science), as well as on questions about the nature and reality of race.

Simon Lohse is Assistant Professor of Philosophy of Science at Radboud University in the Netherlands. He obtained his PhD from Leibniz University Hannover, Germany, was a visiting PhD researcher at the LSE, UK (2013), and held postdoc positions in Hannover (2018–2020) and Lübeck, Germany (2021). His research focuses on pluralism in evidence-based policy and on questions at the intersection of philosophy, ethics, and social studies of biomedicine (esp. in precision medicine and animal experimentation). Simon is currently working on a book project on the philosophy of scientific policy advice.

Nemi Pelgrom is a PhD student in computer science at the Data Intensive Applications group at Linnaeus University. She has a background in the philosophy of science, and her current research is in applications of AI and lowering thresholds for interdisciplinary research.

Thomas A.C. Reydon is Professor of Philosophy of Science and Technology in the Institute of Philosophy as well as the Centre for Ethics and Law in the Life Sciences (CELLS) at Leibniz Universität Hannover. He is also Associated Faculty in the Socially Engaged Philosophy of Science Group (SEPOS) at Michigan State University, an elected Fellow of The Linnean Society of London, one of three Editors in Chief of the book series *History, Philosophy and Theory of the Life Sciences*, and one of three Editors in Chief of the *Journal for General Philosophy of Science*. His current research focuses on the structure of evolutionary explanations, applications of evolutionary thinking outside the biological sciences, philosophical aspects of biological taxonomy, and some issues in research ethics. He recently co-edited the book *Evolutionary Thinking Across Disciplines: Problems and Perspectives in Generalized Darwinism* (Springer, 2023) and authored *The Scope of Evolutionary Thinking* (Cambridge University Press, 2025). For more details about his academic work, see www.reydon.info.

Marc Slors is Professor of Philosophy of Mind and Cognition at Radboud University, Nijmegen. He has worked on topics such as social cognition, free will, and consciousness. His current research focuses on culture as an ingredient, rather than merely a product, of cognition, and focuses specifically on the role of stylistic conventions in social coordination. He has authored two academic monographs, two books for a non-academic audience, and more than 80 papers and book chapters.

J.W. (Jan-Willem) Stoelhorst is Professor of Strategy and Organization at the University of Amsterdam. He is a management scholar with interests in evolutionary theory, economic theory, and behavioral research, and he works on the theoretical and behavioral foundations of (evolutionary, organizational, and institutional) economics and strategic management. The underlying ambition of his work is to turn the study of economic phenomena, in particular the management of firms, into an evolutionary social science.

Martina Valković is a visiting researcher at Leibniz University Hannover and Radboud University Nijmegen. She obtained her PhD from Leibniz University Hannover. Her research has overlapped with topics in epistemology, philosophy of science, social ontology, and social and political philosophy. In her PhD dissertation, she took a critical look at some of the cultural evolutionary theories most dominant today and showed how their ontological and methodological assumptions limit their scope and use, and can lead to negative social and political consequences. She is also active in public philosophy, currently serving as the Public Philosophy beat editor at the *Blog of the APA*.

Acknowledgments

We are indebted to the contributors to this volume for commenting on each other's chapters and to two anonymous reviewers for the press for helpful comments. This volume originated in the workshop "Social Ontology Meets Cultural Evolution", which was held at Leibniz University Hannover in December 2023. We thank Philippe Huneman for suggesting the idea for this workshop to us. The workshop and work on the present volume were supported by a joint grant from the *Deutsche Forschungsgemeinschaft* (DFG) and the *Agence Nationale de la Recherche* (ANR) for the German–French research project "The Explanatory Scope of Generalized Darwinism: Towards Criteria for Evolutionary Explanations Outside Biology", which was led jointly by Philippe Huneman and Thomas A.C. Reydon (DFG project nr. 430626147).

1 Introduction

Building Bridges Between Fields

Martina Valković and Thomas A.C. Reydon

1.1 Why Cultural Evolution and Social Ontology Should Connect

In recent decades, cultural evolutionary approaches to understanding and explaining social and cultural phenomena have flourished, and evolutionary explanations of how human cultures and societies came to be what they are have proliferated (see, e.g., Cavalli-Sforza & Feldman, 1981; Boyd & Richerson, 1985, 2005; Bowles & Gintis, 2011; Mesoudi, 2011; Lewens, 2015, 2024; Henrich, 2016; Tomlinson, 2018; Acerbi, 2020). The phenomena that cultural evolutionists attempt to account for vary widely and range from the production and use of tools and technologies to the origin and change of languages, the biological bases of norms, and the structures of social institutions. In this respect, cultural evolutionists have long promised to provide an encompassing view of the origin and change of human culture and society. Even though cultural evolution as such is not a unified discipline or theory, but rather a loose collection of approaches to explaining human culture and sociality that is unified only by a general evolutionary perspective (Reydon, 2025), cultural evolution has been hailed by some authors as providing a uniquely suited framework to explain the whole of culture and unify the social sciences (Mesoudi, 2011; Mesoudi et al., 2006).

But while cultural evolutionists seek to describe and explain the cultural and social world, they largely take a major aspect of the phenomenon they want to describe and explain for granted: the *ontology* of the social world. Accounts of the evolution of specific aspects of human cultures and societies tend to start from the assumption of the world as populated by individual humans and groups of humans who serve as the bearers of cultural traits. Cultural traits, in turn, are taken to be clearly individuated units of information that are transmitted between people and groups, and have different degrees of success in being transmitted due to different selection pressures operating in the society in which they spread. Cultural evolutionists

DOI: 10.4324/9781003643845-1

think of cultural traits as "ways of doing things that exist in identifiable form over extended periods of time" (O'Brien et al., 2010: 3797), such as specific ways of preparing food or building dwellings, that can be distinguished between cultures and are passed on between generations. Cultural evolution thus (often tacitly) rests on the assumption of a particular strain of individualism, that is, a view of societies as fundamentally consisting of and determined by individual human beings who transmit and copy well-individuated units of culture.

By assuming such an individualistic ontology of the social world to simply be in place as the substrate for cultural evolutionary processes, cultural evolutionists not only assume part of what they should explain, but also pass over core issues and questions that social ontologists have long been concerned with and that, we contend, are crucial issues for cultural evolution research, too. One is the question what – to use an evocative term from Bunge (1977) – the "furniture" of the social world consists of: can societies be completely understood as aggregates of human individuals interacting with each other, or should they be seen as more complex systems in which various kinds of entities play their parts, and if so, what kinds of entities should be recognized (not just people and groups of people, but also institutions such as companies, local and national governments, federations, trade unions, NGOs, etc.)? If the social world is made up of various kinds of entities that interact with each other in complex ways, a number of questions present themselves that do not arise in individualistic approaches.

For instance, if next to people, institutions should also be taken into account as actors in society, what is the nature of institutions, groups, and other collective agents – what makes them act as wholes? Can groups be agents in a similar sense as individual human beings (List & Pettit, 2013)? Do they play similar roles in cultural evolutionary processes as individual humans, or do they not feature at all in such processes? And how do such higher-level actors cause social phenomena? If societies should be understood as systems that are composed of individual humans and various other kinds of parts, rather than as mere aggregates of individuals, an additional question is how the structures of such systems affect the cultural evolutionary processes that occur in them. That is, what is the causal role of societal structures, such as power structures and differences in access to opportunities based on socio-economic status, in shaping how the parts of systems interact with each other, how they change as a consequence of structural factors, and how, in turn, these structural factors themselves change (Valković, 2025)? These are only a few questions that pose themselves when assuming a rich social ontology that goes beyond views of societies as aggregates of individual humans interacting with each other.

The lack of attention in cultural evolutionary theories for such key issues in social ontology is unfortunate, since it means that many central assumptions remain unexamined. For instance, how does the aforementioned widespread assumption of individualism in cultural evolutionary theory influence the development of the theory and its results? Does it (unjustly) highlight a certain organizational level of the social world while obscuring others that are at least equally important? Does it assume a particular view of evolution by taking cultural selection as fundamentally an individual-level process rather than a process that occurs at multiple levels of organization simultaneously, thus missing important causal aspects of the process? What about other, often implicit, assumptions of the theory, such as that the relevant populations are essentially democratically organized, with no serious structural power relations between institutions and individuals? (For discussion of this issue, see Valković, 2025.) Accordingly, one aim of this collection is to examine in what ways social ontology can be relevant for the work done by cultural evolutionary theorists and how cultural evolutionary theory can benefit from a closer engagement with social ontology. We contend that one important benefit for cultural evolution will be that social ontology can provide critical tests of key assumptions in cultural evolution by assessing the feasibility of the ontological commitments underpinning these assumptions. In addition, showing what the socio-cultural world would have to be like for such key assumptions in cultural evolutionary theory to be valid can help researchers assess how far the evolutionary models they use are removed from social reality (i.e., to what extent their models involve idealization and reduction of the complexity of social reality – for a general account of these issues, see Potochnik, 2017), and to estimate under which conditions and to which kinds of societies the various evolutionary approaches that are available in the cultural and social sciences can be applied.

Conversely, social ontology as a field of study has shown insufficient interest in cultural evolution. If cultural evolutionary theory provides correct explanations of cultural change (which is still open for debate), or if it at least provides reliable models that show how novel cultural traits can, *in principle,* arise in evolutionary processes, it also tells us something about the (possible) origins of the various kinds of cultural and social entities we encounter in the world. That is, cultural evolution could provide important information about the origins of social entities at various levels of organization, the ways in which these change and become adapted to their environments, and their adaptive potential with respect to future environmental changes.

In addition, a cultural evolutionary perspective may provide clues regarding the "furniture" of the social world: from an evolutionary viewpoint, for institutions and other kinds of social entities to be counted as actors in social contexts rather than mere aggregates of people, arguably they should have an evolutionary history of their own. Cultural evolution can – at least in principle – have given rise to higher-level social agents in the same way as biological evolution produced living beings with various degrees of agency (or precursor traits) with humans as paradigmatic agents (Tomasello, 2022). Whether or not cultural evolution *did* produce higher-level social agents is a topic for both empirical and conceptual research that should show how higher-level agency may have evolved on the basis of human agency (i.e., as a series of next steps in the evolution of agency) and clarify what agency *is* in the first place (which remains a debated issue in biology – see, e.g., Okasha, 2018, 2024; DiFrisco & Gawne, 2025; Fábregas-Tejeda et al., 2025). Cultural evolutionary history thus could provide a basis for assessing the ontological status of putative social entities as full-blown social actors rather than merely constituting the environment for acting human beings. By illuminating the possible evolutionary origins of social entities, cultural evolution can provide a basis for assessing the feasibility of the social ontologies that researchers in social science projects assume. Accordingly, the second aim of this volume is to inquire how social ontology could benefit from engaging with cultural evolutionary research.

In this volume, we offer a remedy for the current lack of engagement between the two fields by bringing cultural evolution and social ontology into a dialogue. We aim to show that cultural evolution and social ontology are complementary disciplines that, while each has their own subject matter, research questions, and methodologies, rely on each other's results, both by mutually restricting each other and by opening up new perspectives. In so doing, we intend this volume as a first pass at showing how cultural evolution and social ontology can complement each other, and in this way as a starting point for the development of closer connections between the two fields from which both will benefit. We will now first turn to the current state of the art in cultural evolution and social ontology as well as the history of connections between these disciplines, and then, in the remainder of this chapter, provide a brief overview of the structure of the present volume.

1.2 Contextualizing the Volume: Where We Are and How We Got Here

Most of the existing literature on the topic of the present volume has focused either on the field of cultural evolution or social ontology, but not on both these fields in dialogue. For instance, some of the most prominent recent works in cultural evolution – Alex Mesoudi's *Cultural Evolution: How Darwinian Theory Can Explain Human Culture and Synthesize the*

Social Sciences (Mesoudi, 2011); Tim Lewens' *Cultural Evolution: Conceptual Challenges* (Lewens, 2015); and Kevin McCaffree's *Cultural Evolution: The Empirical and Theoretical Landscape* (McCaffree, 2022) – do not even mention (social) ontology.[1] The same holds for Lewens' recent *Cultural Selection* (Lewens, 2024), W.G. Runciman's *The Theory of Cultural and Social Selection* (Runciman, 2009), Marion Blute's *Darwinian Sociocultural Evolution: Solutions to Dilemmas in Cultural and Social Theory* (Blute, 2010), Olivier Morin's *How Traditions Live and Die* (Morin, 2016), Gary Tomlinson's *Culture and the Course of Human Evolution* (Tomlinson, 2018), and Alberto Acerbi's *Cultural Evolution in the Digital Age* (Acerbi, 2020). Only Lewens (2015) here and there touches upon presumably ontological questions, such as whether culture is to be seen as information or populations are to be understood as collections.

This is not to say that these works do not address ontological aspects of human cultures and societies – all of cultural evolution is concerned with aspects of human cultures and societies that come with ontological questions. The evolution of languages, religion, and social structures generally all involve the need to address at least *some* ontological issues (how should languages or religions be classified to make phylogenetic studies possible, for instance). But the point is that ontology remains in the background, that ontological questions are only considered implicitly, and that virtually no connections are forged with work in the field of social ontology. Moreover, fundamental issues that are specific for ontological research, such as the nature of the various kinds of social and cultural entities (are social kinds a subcategory of natural kinds, are they human constructions, or something in between?), the foundations of classification of social kinds and entities (should such entities be classified on the basis of similarities in shape and structure, on the basis of function, on the basis of descent, or on a mixture of criteria?), the question what the ontology of the social world consists of (objects, properties, dispositions, structures, etc.), and so on are not being addressed. In this sense, even though all cultural evolution touches upon some social ontological issues, the fundamental questions of ontology remain outside the scope of concern. An explicit and systematic treatment of social ontological issues is lacking in this literature.

In recent social ontological literature, there is a little more engagement with evolution. Brian Epstein's *The Ant Trap: Rebuilding the Foundations of the Social Sciences* (Epstein, 2015) mentions evolution in passing but does not develop on it or engage with cultural evolutionary theories. Epstein's recent *Social Ontology* (Epstein, 2025) does not mention evolution at all. Raimo Tuomela's *Social Ontology: Collective Intentionality and Group Agents* (Tuomela, 2013) is possibly the work that combines social ontology and cultural evolution the most. It takes into account the evolution of humans, acknowledges gene–culture coevolution, and draws attention to how the circumstances in which our ancestors lived could have led to the evolution of certain ways of thinking. However, Tuomela does not consider these topics in depth or engage with cultural evolutionary literature, as his focus remains firmly elsewhere. Similarly, Francesco Guala's *Understanding Institutions: The Science and Philosophy of Living Together* (Guala, 2016) acknowledges human evolution in passing, but apart from some sections on evolutionary game theory and metaphorical talk of "evolution of institutions", does not engage with the topic or with cultural evolutionary literature.

In brief, there is a lack of available literature focused firmly on the intersection of social ontology and cultural evolution, and that explicitly connects these two fields of research. Social science and evolutionary theory have a long history of connectedness, tracing back at least as far as Herbert Spencer's work in the late 19th century (Spencer, 1890, 1891). Darwin, too, was concerned with understanding central aspects of human sociality from an evolutionary perspective. But while ontological questions have always to some extent been part of evolutionary sociological research, they have typically remained in the background, and explicit attention to social ontology in evolutionary sociology has been very limited. In part, this may be because social ontology has always been more a subfield of philosophy than of social science (in the same way as, for instance, ontological questions about the physical and biological world are addressed much more by philosophers of physics and philosophers of biology than by physicists and biologists themselves).

Moreover, as Brian Epstein (2025: 1) points out, while social ontology can be seen as one of the oldest fields in philosophy, it has not always been *en vogue*. In particular due to the Vienna Circle's strong opposition to metaphysics, interest for metaphysical and ontological issues strongly declined among philosophers in the first half of the 20th century – that is, precisely at the time when Darwinian evolutionary theory again gained dominance after the "eclipse of Darwinism" from roughly the 1880s through the 1920s (Bowler, 1983; Reydon, 2025: 22ff.). During this "eclipse" period, biologists generally accepted *that* evolution occurred, but were much less in agreement about *how* it occurred. In particular, Darwin's theory of natural selection was not generally accepted, as many biologists thought there was insufficient evidence to support it, and many authors formulated alternative theories of evolution, sometimes presenting them as strongly anti-Darwinian theories. Darwin's ideas returned to the center of evolutionary thinking only with the establishment of the Modern Synthesis in the 1930s–1940s, that is, at a time when metaphysics generally was not a central topic in philosophy. We suggest (and admit this suggestion to be speculative, as we are not historians of the social sciences) that this may be a reason for the disconnect between evolutionary theorizing and social ontology.

Let us speculate some more. Social ontology as a recognizable subfield of philosophy established itself only comparatively recently: the first issue of the *Journal of Social Ontology* appeared in 2015, and the *International Social Ontology Society* was founded in 2012 (as successor of the *European Network on Social Ontology* from 2007).[2] Only today, "the field is beginning to coalesce" (Epstein, 2025: 1), but social ontology as a field is still characterized by radically diverging approaches and perspectives (Epstein, 2025: 1) as it "brings together theoretical work from a large number of disciplines" (Schmid et al., 2015: v). Moreover, there is some difference in opinion regarding the scope of social ontology. Epstein, for example, defines the field as broadly as possible:

> Social ontology is the study of the nature of the social world. "What is X?" social ontology asks, where "X" stands for *any social thing whatever*. What, for instance, are groups? What are corporations? What is law? What is race? What is gender? What is language? Fill in a social object, a social kind, a social category – and it's a topic for social ontology. Even more general questions are in its purview as well: What is society? What is the social?
>
> (Epstein, 2025: 1, emphasis added)

Lynne Rudder Baker, in contrast, sees things differently. In a paper that she submitted in late 2017, shortly before her passing, and that is titled "Just what is social ontology?", she writes:

> Social ontology is a budding branch of philosophy, but there has been little consideration of what exactly social ontology is. Social phenomena include financial transactions, cocktail parties, two people walking together, athletic teams, and legislatures, among countless other things – a "motley crew", as Margaret Gilbert has called them. . . . Construing ontology as an inventory of what genuinely and nonredundantly exists, do all – or any – social phenomena belong in ontology?
>
> (Baker, 2019: 2)

While both Epstein and Baker point out that social ontology is a young field, Epstein sees all "What is X" questions about aspects of human societies and cultures as falling into the purview of social ontology, whereas for Baker, social ontology is about *nonredundant* aspects of human societies and cultures. Baker's view is reminiscent of Quine's (1948) preference for "desert landscapes", that is, the methodological principle to only accept in one's ontology entities and kinds that cannot be reduced to other, more fundamental entities and kinds. On Baker's view, it makes sense to ask whether cocktail parties and two people walking together should be recognized as genuine parts of social ontology, or whether these are redundant parts that could be reduced to more fundamental aspects of society and culture (such as individual people and transmitted behaviors, or more general practices like the practice of building social networks or the practice of consuming consciousness-altering and sociability-enhancing substances).

Baker holds that

> Ontology . . . comprises all the nonredundant and irreducible objects, properties and kinds required to make sense of the world. . . . Social ontology, on my view, is that part of a nonredundant inventory of reality that includes social individuals, properties and kinds.
>
> (Baker, 2019: 11–12)

On such an approach, possible connections to cultural evolution are restricted by what social ontologists acknowledge as parts of social ontology. If cocktail parties are accepted as nonredundant and irreducible parts of social ontology, it makes sense to inquire into the evolutionary history of cocktail parties. If they are not, the cultural evolution of cocktail parties is also not something that social ontologists will deem relevant. The point is that the need to make connections with cultural evolution from the side of social ontology depends on how social ontologists conceive of their own field. And as social ontology is a young field that encompasses a variety of approaches, it should not be surprising that social ontologists from their side have not been building many bridges to cultural evolution.

And as was already mentioned, something similar holds for cultural evolution. Cultural evolution is not a unified discipline, nor do researchers converge on a single theory of cultural evolution, but it is a collection of – sometimes strongly diverging – approaches to explaining a large diversity of aspects of human culture and social interaction. As such, cultural evolution overlaps with a considerable number of fields and subfields that are spread out over diverse corners of academia, including evolutionary anthropology, evolutionary literary studies, evolutionary sociology, evolutionary economics, evolutionary studies of science and technology, and more. Here, we do not only find diverging conceptualizations of what evolution *is* and what makes a process of change into an *evolutionary* process, as well as diverging ways of connecting to evolutionary biology (Reydon, 2025) – we also find evolutionary approaches embedded in diverging ways of doing research. Economists do not conduct research in the same way as anthropologists do, and neither work in the same way as philosophers of science and technology do, who try to understand scientific and technological changes as evolutionary processes. The diverging traditions of doing research that can be found in the different corners of academia affect the way in which theories and data are interpreted, in which concepts are understood, and the way in which bridges to other fields can be built.

In light of this brief sketch, perhaps it should not be surprising that explicit connections between social ontology and cultural evolution have been scant. Moreover, those connections that have been made are not always the kinds of connections that fit contemporary views of what cultural evolution and social ontology encompass.

While the intersection of cultural evolution and social ontology has not been a focus of much, if any, explicit interest, there is literature tracing back to the mid-20th century that to some extent interacts with both of these fields. Philosopher and anthropologist David Bidney's book, *Theoretical Anthropology* (Bidney, 1953), for instance, discusses concepts in social science and anthropology from a cultural evolutionary perspective. Bidney's interests, however, lie primarily with the progressive change in human cultures through the unfolding of human potential, and not so much with the nature of social entities and kinds. Ontological questions remain largely in the background of his work, and his work does not explicitly connect social ontology and evolution. Moreover, Bidney's view of evolution diverges from contemporary accounts of cultural evolution. Bidney sees cultural evolution as driven by human efforts at achieving the potential of human nature: "Cultural evolution may be understood as the *unfolding or activating of the potentialities of human nature* through a process of self-conditioning and education in relation to a given environment" (Bidney, 1953: 82; emphasis added). Accordingly, Bidney holds that in the sphere of culture, Lamarckian evolutionary theory is more adequate than Darwinian evolutionary theory, as "Darwinian evolution allows no place for human effort, and when applied to the human situation it leads to a kind of cultural fatalism" (Bidney, 1953: 82). Bidney's view of evolution as a process of unfolding features of a system that are already present as potentialities in the system indeed aligns more with Lamarck's or Herbert Spencer's views of evolution than with Charles Darwin's view or with contemporary evolutionary theory. In particular, it resonates with Lamarck's and Spencer's views of evolution as a fundamentally directional, progressive process – aspects that disconnect Bidney's understanding of evolution from contemporary approaches in cultural evolution.

Around the same time as Bidney, anthropologist Julian Steward published his *Theory of Culture Change: The Methodology of Multilinear Evolution* (Steward, 1955), in which he put forward his evolutionary theory of cultural change and reflected on the concepts and methods employed in his approach. Somewhat later, anthropologist Leslie White published one of the major works advocating for a cultural evolutionary approach in studying how societies and cultures develop: *The Evolution of Culture* (White, 1959). Steward's and White's books were seminal works that ushered in what is usually described as a neo-evolutionary school of anthropology (even though Steward strongly disagreed with White's approach – see below). But in many ways, this neo-evolutionary school was quite different from contemporary cultural evolutionary approaches. It was a reaction against earlier views of societal and cultural evolution as a linear progression from more primitive to more advanced stages that all societies and cultures passed through, such as Herbert Spencer's influential view that all societies evolve progressively through the same four stages (Weinstein, 2024). And in Steward's and White's works, too, social ontology is not an explicit topic but remains in the background, even though there is some concern with the nature of societies.

The conception of evolution in the background of Steward's and White's views was a very broad understanding of evolution as non-cyclical, irreversible change that – even though these authors were influenced by Darwin's ideas – did not connect much to biological evolutionary theory. Where they did connect, they understood evolution in a different way from contemporary cultural evolutionary theorists. In particular, they did not explicitly distinguish between population-level evolution and organismal development (a distinction that is central in biological theory) and were equally willing to describe societies as developing organisms or as evolving systems. This perspective can be traced back to Herbert Spencer, who also did not strictly distinguish between evolution and development, and wrote about society as a developing organism. The neo-evolutionary school of anthropology retained this perspective, placing itself at some distance from biological evolutionary thinking.

Steward, for instance, aims to establish a field of cultural *ecology* rather than cultural evolution. In his view, cultures should be understood as developing not in isolation by way of a process that progresses linearly through the same stages (as earlier theorists thought) but rather in interaction with their specific natural environments. In Steward's view, cultures thus develop in different ways, and environmental factors (local climate, availability of various kinds of resources, aspects of the local landscape, etc.) are seen as important factors that affect how a culture develops. Still, Steward's account of cultural change was not an evolutionary account in the sense of contemporary cultural evolution or biological evolutionary theory. For example, he describes his approach as follows:

> Cultural evolution . . . may be defined broadly as a quest for cultural regularities or laws. . . . Multilinear evolution is essentially a methodology based on the assumption that significant regularities in cultural change occur, and it is concerned with the determination of cultural laws. Its method is empirical rather than deductive. . . . Multilinear evolution . . . has no a priori scheme or laws. It . . . simply

> poses the question of whether any genuine or meaningful similarities between certain cultures exist and whether these lend themselves to formulation.
>
> (Steward, 1955: 14, 18–19)

Steward's approach (multilinear evolution) thus does not assume specific theoretical components of Darwinian evolutionary theory, but rather aims to derive regularities of cultural change from empirical research. This is a bottom-up approach that, if successful, would yield laws of cultural "evolution", but these would not be the same laws that apply to biological evolution. In a later paper in the prominent biological journal *Evolution*, Steward discusses "the biological factor in cultural evolution" (Steward, 1958: 206) and explains that this encompasses "the possible ways in which inherent physical, physiological and psychological characteristics shared by all human beings have conditioned culture and channeled the direction of its development" (Steward, 1958: 206). For Steward, the fundamental problem of cultural evolution is

> how these factors have imposed limits on the range of cultural variation, how they have channeled the directions of the evolution of any culture and how they have interacted with cultural forms to give the latter special function and meaning.
>
> (Steward, 1958: 210)

This shows that Steward is concerned with the evolutionary *history* of humans in relation to culture, but not with cultural and societal changes as evolutionary *processes* themselves (cf. Reydon, 2025: Section 4.1). Steward's work is inspired by evolutionary ideas (in the sense that he sees cultures as evolving or developing) and uses results from evolutionary research, but it does not use the theoretical framework of evolutionary biology and does not take up biological evolutionary theory in the way contemporary cultural evolutionists do.

In this respect, Steward's view of cultural evolution is still closer to contemporary biology than White's. In his book, Steward examines three different interpretations of evolutionary thinking in relation to cultural change (Steward, 1955: 14–19), none of which, however, uses the explanatory framework of contemporary biological evolutionary theory, or Darwin's theory, and explicitly positions his approach in contrast to White's (with whom Steward strongly disagrees). Steward places White's views in the category of views that see all cultures as developing or evolving through the same stages and interpret this change as progressive, that is, as assuming a universal process of cultural and societal development or evolution.

White explicitly addresses questions that today would fall in the field of social ontology in chapters on the nature of social organization and on the structure, functions, and evolution of social systems. But his evolutionary approach, too, is detached from biological evolutionary theory. White defines evolution in a very general way as gradual and irreversible change, without explicitly incorporating biological ideas. He does not distinguish between evolution and development and writes:

> Culture undergoes another kind of change, also: a developmental or evolutionist process of change. Evolution may be defined as a temporal sequence of forms: one form grows out of another, culture advances from one stage to another . . . a mere aggregation of things without *organic unity* cannot undergo evolution.
>
> (White, 1959: 29–30; emphasis added)[3]

He conceives of cultures and societies as superorganismal entities that develop or evolve according to their own laws and are driven by greater efficiency in harnessing energy through technology. In an earlier paper, he writes that

> Culture is a kind of behavior. And behavior, whether of man, mule, plant, comet or molecule, may be treated as a manifestation of energy. Thus we see, on all levels of reality, that phenomena lend themselves to description and interpretation in terms of energy.
>
> (White, 1943: 335)

Applied to cultural and societal change, White asserts that changes as due to

> a never-ending attempt to make of culture a more effective instrument with which to provide security of life and survival of the species. And one of the ways of making culture a more powerful instrument is to harness and put to work within it more energy per capita per year. . . . The other way of improving culture as an instrument of adjustment and control is to invent new and better tools and to improve old ones. Thus energy for culture-living and culture-building is augmented in quantity, is expended more efficiently, and culture advances. Thus we know, not only *how* culture evolves, but *why* as well.
>
> (White, 1943: 339; emphasis in original)

White thus attempts to devise an overarching account of behavior and behavioral change in terms of energy use that could apply in all areas of the natural, life, and social sciences. As in Bidney's view, human intentionality is the central driving force of cultural evolution, but contrary to Bidney, it is not about unfolding the inner potential of human nature, but about the more efficient harnessing of energy by means of better technology. Here, too, we see an approach to cultural change that is inspired to some extent by evolutionary thinking, but falls far short of constituting a genuinely evolutionary theory of culture. White assumes a specific causal framework to explain cultural and societal change, but his framework is very different from that of contemporary evolutionary theory.

This very brief and very sketchy historical overview has highlighted several reasons why relevant connections between cultural evolution and social ontology have been rare. Of these, we believe that a very important reason is related to the availability of divergent understanding of what evolution is and how it occurs. Biologists themselves have debated these questions ever since Darwin published his theory, and researchers in areas outside biology have taken up a variety of such views (see Reydon, 2025). In particular, early cultural evolutionists such as David Bidney, Julian Steward, and Leslie White held views of evolution that do not fit well with biological theory and did not explicitly take up a Darwinian framework. In addition, while they were implicitly concerned with social ontological questions, they did not connect to the field of social ontology, as that still had to (re-)emerge as a subfield of philosophy.

Recent work from the 1980s onward in several fields, including evolutionary biology, the philosophy of biology, cultural evolution, and social ontology, more explicitly addresses connections between cultural evolution (with views of evolution that align more closely with biological theory than in earlier work) and social ontology.

In biology, for example, Maynard Smith and Szathmáry's pioneering work, *The Major Transitions in Evolution* (Maynard Smith & Szathmáry, 1995), discussed the major changes in information organization and transmission, from the origin of life and first cells, to the emergence of societies and language. The common feature of these transitions, Maynard Smith and Szathmáry claimed, is that the entities on lower levels lose the ability to reproduce, becoming able to do so only on the upper level. Maynard Smith's and Szathmáry's book contains two chapters on the evolutionary origins of societies and of language, and forges some connections to ontological issues (but not to fundamental issues in ontology, such as the nature of social and cultural kinds, or the question regarding the fundamental ontology of the social world). Boomsma (2023) provides a treatment of the major transitions that is more focused on social evolution.

Currently, a growing interest in this topic is seen, and the literature that connects cultural evolution and social ontology is growing rapidly (e.g., Hodgson & Knudsen, 2006; Aldrich et al., 2008; several chapters in Krohs & Kroes, 2009; Stoelhorst, 2014; O'Connor, 2019; Valković, 2024;

Buskell, 2025). A special section of the journal *Topoi* (vol. 35, issue 1, 2016) brings together work from the two fields. Rapid advances in genetics and genomics in the past decades made it possible to integrate evidence from genetics and genomics into research on ancient cultures and societies (e.g., Hiscock & Sterelny, 2025). A recent approach aiming to synthesize elements from recent developments in biological evolutionary theory and aspects from human sociality is that of Joseph Rouse (2023), who presents an account of social practices as instances of niche construction, thereby advocating for an evolutionary explanation of one of the basic social-ontological concepts. Still, even though cultural evolution and social ontology gradually become more intertwined, to a large extent the two fields continue to lead separate lives and accordingly, we believe, there is a need for building more bridges, which the present volume aims to do.

1.3 Structure of This Volume

Against the background of this state of affairs, the novel contribution of this volume is in exploring how methods, approaches, and concepts of social ontology, and those of cultural evolution, can cross-fertilize. This is a much-needed contribution to both the philosophy of social science and cultural evolution. For this purpose, we bring together contributions from experts in the two fields, whose expertise reaches into related fields such as social philosophy, philosophy of language, philosophy of biology, philosophy of the social sciences more broadly, and philosophical methodology. While the authors vary in their approach to the relationship between social ontology and cultural evolution, there are many common threads in the nine main chapters of the volume: chapters focus on the themes of power, language, agency, interaction, and social institutions and roles. While the contributors to this volume work either in the area of social ontology or cultural evolution, all chapters construct a bridge to the other area as well, as we will show below.

Besides researchers working in the two focal areas of this volume, that is, cultural evolutionists and social ontologists, the discussions presented in this volume, perhaps in particular as they include analyses of various forms of power, institutions, and collectives, should be of relevance to researchers in political philosophy, philosophy of the social sciences, philosophy of biology, anthropology, and sociology, as well as philosophers and social scientists more generally. The chapters are divided into three thematic sections, each comprising three chapters: *metaperspectives* on cultural evolution and social ontology (Chapters 2–4), the *power* concept in cultural evolution and social ontology (Chapters 5–7), and *interactions and collectives* in cultural and social evolution (Chapters 8–10).

The first three chapters all assume a certain metaperspective to reflect on the relationship between social ontology and cultural evolution. The volume opens with a chapter by Lohse and Valković (Chapter 2), who turn to mapping critical arguments that have been raised against cultural evolutionary approaches and the connections between them, in order to better understand the prospects and limitations of these approaches (and their criticism) and to illustrate how social ontology can contribute to the critique of evolutionary approaches, especially when combined with epistemological and political arguments. They focus on arguments targeting what they claim to be three tendencies of evolutionary approaches to societies and cultures: (1) relying on a research strategy of reductive individualism, (2) a reductive or minimizing approach to human agency, and (3) population-based modeling of socio-cultural entities. Lohse and Valković discuss arguments critical of these assumptions by dividing them into ontological, epistemological, and political ones, while stressing the relations between them. In this way, their chapter directly relates to the long-standing debates in social ontology regarding reductive (methodological) individualism, human agency, and the nature of reproducing entities in the ontology of cultural evolutionary theories. They make a case not only for the interrelation of the three aspects of criticism (ontological, epistemological, and political), but also for the relations between the three criticized tendencies, arguing that the first two are instances of reductionism, while the first and the third ones can, curiously, be seen as opposed in a sense. This, they claim, makes it clear both that there might be a certain worldview behind these cultural evolutionary approaches and that any criticism of cultural evolution must pay attention to the specific theoretical commitments, goals, and background assumptions of the particular evolutionary approach it aims to target.

In the following chapter (Chapter 3), Lauer argues for a method of pursuing projects in social ontology that would rest on empirical results of the social sciences inferred by using the most discerning available methods. He distinguishes between two kinds of inquiry in cultural evolution studies: (1) modeling-based studies, the goal of which is to provide an abstract model describing the possible emergence of some social practice; and (2) empirical studies, the goal of which is to find a way of testing hypotheses about the emergence of a social practice from operationalized concepts. If we want to know what the social world is like, Lauer argues, we must rely on the results of empirical studies when pursuing ontological projects. This chapter connects studies of cultural evolution to the metaphilosophical debate on how to best pursue ontological projects, illustrating how cultural evolutionary studies could contribute to the metaphysical work on the nature of social reality. It also introduces an account of "discerningness" as a constraint on empirical studies from which we should draw upon when discussing the ontology of the social world, thereby offering an approach that is in tension with broader literature, including other chapters of this volume (Chapters 5–7). Chapter 3 also connects to the previous chapter in the volume in that Lauer, on the basis of the kinds of inferences we can draw from cultural evolutionary empirical models, argues that ontological individualism can be rejected. In this way, the chapter also relates to the wider debates on naturalism and reductive individualism in social ontology, showing how the results of cultural evolutionary inquiries can be of relevance for the debate.

Hindriks (Chapter 4) argues that rational choice functionalism, one of the few non-evolutionary functionalisms still defended in the social sciences, faces what he terms the axiological problem: since it relies on value judgments to distinguish between institutions that have or do not have a function, it is inadequate as an explanatory theory. Evolutionary functionalism that plays a role in cultural evolutionary theory, on the other hand, faces what Hindriks calls the complexity problem: it fails to capture the complexity of institutions because it identifies culture with traits of individuals. To solve these problems, Hindriks argues we need a richer theory of institutions that will involve rules that prescribe actions or give them meaning. Rules, according to him, provide a non-arbitrary way of identifying institutions that have a function and those that do not, without recourse to values. They also capture institutions' symbolic and normative features and can account for social roles and material resources. Hindriks uses this theory of institutions to show that we can develop two conceptions of social functions, evolutionary and non-evolutionary, the first of which would account for the reproduction, and the second of which would account for the persistence of institutions, and which would face neither the axiological nor the complexity problem. In this way, he shows how functionalisms in the social sciences can address these two problems. By identifying the complexity problem of cultural evolutionary theory, ontological and methodological individualism, and holism, Hindriks gives his own contribution to the debate already elaborated on in the previous two chapters of this section, focusing on the second version of the problem Lohse and Valković discussed, as well as on the long-standing debates in social ontology and cultural evolution. It systematically explores social ontology and the methodology of cultural evolution to show how, as it stands, the latter does not capture the complexity of cultural phenomena. The chapter contributes to the study of functionalism in social ontology and philosophy of social science. The chapter's discussion of social roles and power foreshadows the same discussion from Baraghith's and Jordan's chapter (even though the latter do not employ the concept of a rule, but that of a niche, in their account of social roles), as well as the discussion on teleological conception of normativity from Chapter 7. The chapter contributes to the cultural evolution studies with the notion of methodological pluralism, which could be used to avoid some of the most prominent problems in the field.

The focus of the next three chapters of the volume is power, variously construed. Starting from the premise that power is a central feature of our societies and a fundamental part of our social and cultural worlds, Chellappoo (Chapter 5) revisits Richard Lewontin's critique of cultural evolution as unable to adequately explain power due to its commitment to methodological individualism. She analyses several of the ways in which different cultural evolutionary approaches tried to account for social groups and power in individualistic terms, suggesting that race and gender hierarchies are among the structural power relations that at least some of these frameworks have difficulty accommodating. While she argues that none of the strategies of reduction, restriction, presumption, nor the cultural niche construction approach (as exemplified by Baraghith and Jordan in this volume) present us with satisfactory solutions, she points to a growing body of work in evolutionary game theory, where these models are used to provide some insights into conditions under which gender and racial hierarchies could emerge. Chellappoo argues that the success of these models when it comes to generating insights into structural inequality somewhat weakens Lewontin's critique. Recent developments in the evolutionary game theoretic work challenge the claim that individualist commitments prevent cultural evolutionary theories from accommodating power. However, Chellappoo highlights that these approaches remain incomplete, and holds that Lewontin's critique remains powerful in its analysis of problems of certain kinds of cultural evolutionary theories and sets of assumptions, even though an individualist approach could be more successful when it comes to gaining insights into structural phenomena than Lewontin anticipated. Chellappoo's chapter brings together topics from social ontology and methodology, discussing the connections between individuals and higher-level entities such as social structures and groups. It shows how social ontology debates could inform work in cultural evolution by potentially weakening the basis for methodological individualism that is at the center of much cultural evolutionary work, thereby leading to a reassessment of those theories' ambitions, especially in areas such as complex institutions and higher-level structures.

Baraghith and Jordan (Chapter 6) take a different approach, with an aim to integrate social structuralist elements into cultural evolutionary theory by presenting coercion as a form of cultural niche construction where agents inhabit and create social asymmetries through their interactions. They think of power as a social asymmetry that can appear in various forms, including that of prestige, coercion, or dominance. In their view, prestige and coercion are distinct selective strategies, the first of which can be presented in the evolutionary ecology terms as r-strategic behavior, while the second one they propose to see as K-strategic behavior, which would explain its not being frequently replicated yet still exerting influence. Baraghith's and Jordan's chapter has important implications for how we understand cultural evolution: on their analysis, cultural traits are not merely passed on, but rather actively selected based on the social and environmental conditions that are, to a significant extent, influenced by power structures, which can hinder or promote the spread of these traits. Their chapter also contributes to social ontology and philosophy of social science literature on power, with the conception of structural coercion as a form of cultural niche construction. It also calls attention to the limits of current cultural evolutionary theory when it comes to the nature and role of power in human societies, and shows a way forward in integrating this important, but typically overlooked, concept and thereby expanding the boundaries of cultural evolutionary theory. Their chapter, therefore, enriches both the study of social ontology and science as well as that of cultural evolution.

Burman and Pelgrom (Chapter 7) identify the concept of power as central both to social ontology and cultural evolution. With their chapter, they aim to contribute to its conceptual analysis by developing three different notions of social power – telic, artisanal, and deontic – and applying them in an empirical study of the peer-review process. In both social ontology and cultural evolution, Burman and Pelgrom argue, transparent forms of social power have been overemphasized, while opaque forms have been neglected, even though they are central when it comes to understanding societies and their evolution. The authors claim that the existence of opaque forms of power poses a new challenge to cultural evolutionary theory, which it cannot easily accommodate, as this power is non-reproductive in an even stronger sense than coercion, which Baraghith and Jordan aim to integrate. Contemporary social ontology, as well, largely leaves out the opaque forms of social power, Burman and Pelgrom claim. The chapter therefore contributes to both the fields of cultural evolution and social ontology by highlighting this central and understudied feature of our societies.

The last three chapters of the volume center on collectives and interactions between levels in the cultural evolution of coordination and social organization. Geurts (Chapter 8) understands social ontology as an assortment of conventions, norms, values, and institutions, among others, many of which are common ground between members of a society. Common ground is, in his account, a kind of higher-order sharing that includes common knowledge, joint acceptance, mutual belief, and joint commitment, and that is foundational to human society. Common ground is a special kind of correlation device – a circumstance external to an interaction, the main purpose of which, Geurts writes, is to help individuals coordinate their actions by conditionalizing their behavioral strategies. Geurts elaborates on the concept of common ground and contrasts it to that of shared ground, an inclusive notion that includes forms of first- and higher-order sharing that serve as correlation devices. In this way, the concept of a shared ground can be used in the description of interactions as varied as those of bioluminescent bacteria, butterfly "conventions", and human institutions. This framework therefore situates highly complex higher-order phenomena such as norms and institutions on a spectrum of the evolution of coordinated action. Geurts' chapter discusses and employs many central social ontological concepts, such as mutual belief, mutual commitment, higher-order sharing, norms, and institutions, among others, in an account of the evolution of correlation devices and coordinated behavior. Informed by this account, he argues for a concept of institutions as hybrid entities that both create norms and are governed by them, and many of which, such as universities, are also agents. In this, Geurts' account directly relates to that of Stoelhorst in this volume, and the two could be seen as complementary, highlighting different aspects of the same process.

In the following chapter (Chapter 9), Slors illustrates another way in which cultural evolution can contribute to social ontology. He argues that the concept of collective intentionality, which itself was first borrowed from social ontology and applied in the field of cultural evolution, can reenter the domain of social ontology and offer something missing in the literature so far: a "non-mysterious" kind of top-down influence of group-level processes on individuals. Slors argues that cultural group selection can best explain "suprapersonal, transgenerational self-organizing cycles of socialization" that turn individuals into functioning group members who socialize their offspring in a similar way. This, according to Slors, represents a genuine group-to-individual top-down influence which can be useful in the domain of social ontology as well as in that of cultural evolution, implying a bidirectional influence between the two disciplines. This kind of top-down influence, Slors claims, is fundamental in the sense that what is influenced is the very nature of individual-level processes, that is, of how an individual decides or intends. The top-down influence of such a process, Slors argues, becomes apparent once we zoom out and think of this mindshaping as a process encompassing many generations. This is, according to Slors, a group-level, blind evolutionary process. Slors believes it is likely that a parallel to this kind of process exists in the context of collective phenomena such as companies and institutions, posing an opportunity for further research. Slors' chapter invokes long-standing debates in social ontology, such as on levels of causation and top-down causality, individualism and anti-individualism, collective and joint intentionality, and the theory of institutions. It showcases how a notion, that of collective intentionality, can cross from social ontology to cultural evolution and then back again, enriching our understanding in the process. In this way, by taking an evolutionary view on the phenomenon of collective intentionality, social ontological debates are enriched with an account of the formation of individuals as group members.

Stoelhorst (Chapter 10) aims to examine the nature of the institution of the modern firm, arguing that it should be understood as a product of gene–culture coevolution and as a "major transition" in cultural evolution, in the sense of being a form of organization irreducible to the properties of its individual members and an agent in its own right. The chapter combines insights from different evolutionary approaches, including evolutionary epistemology, gene–culture co-evolution, multi-level selection theory, and, most centrally, the theory of major transitions in evolution, outlining how they can be combined into a single framework for studying human social organization, which is then used to analyze the institution of the modern firm as a major transition in cultural evolution. This means that the firm, or the corporation, should be seen as a unit of selection with its own emergent, group-level properties, in the form of legal specifications of social roles and behavioral norms related to them, which get culturally transmitted and define the group's social structure. In his ontology of the firm, roles and norms are what link the individual and collective levels. With his analysis, Stoelhorst contributes to both cultural evolution and social ontology. In the domain of cultural evolution, he demonstrates how the theory of major transitions in evolution can be generalized to cultural evolution, thereby going beyond the individual-level perspective offered by much of the literature (and criticized by several contributors to this volume). Stoelhorst also demonstrates how envisaging a particular kind of institution as a product of gene–culture coevolution can shed light on its ontology, thereby contributing to the work in social ontology on institutions.

Finally, in the last chapter of the volume (Chapter 11), we, the editors of the present volume, explore where more connections are needed and are possible between the areas of cultural evolution and social ontology, and provide some suggestions for successfully building such connections.

The rationale for choosing these particular topics for inclusion in the volume follows from the current state of affairs in cultural evolution and social ontology, as discussed in the previous section. Our selection addresses some of the gaps in the current literature, but of course, we cannot aim for completeness with respect to such a broad topical area. Our aim is to put topics on the agenda for further work that will fill these gaps. As such, the volume is intended as a step toward future research that connects cultural evolution and social ontology more closely, without the presumption of being able to provide an exhaustive coverage of all relevant issues. For that, much additional work will be needed, and we hope to be able to inspire some of it.

Notes

1 McCaffree (2022: 191) mentions "ontology" once, but not in the sense of the philosophical study of ontology. He points out that organisms have ontologies in the sense that every organism has some representation of its environment that allows it to navigate it.
2 See Slade-Caffarel (2022), who also provides some history on Cambridge Social Ontology.
3 White holds that cultures undergo change processes that are different from the diffusion of cultural traits between cultures.

References

Acerbi, A. (2020): *Cultural Evolution in the Digital Age*, Oxford: Oxford University Press.
Aldrich, H.E., Hodgson, G.M., Hull, D.L., Knudsen, T., Mokyr, J. & Vanberg, V.J. (2008): In defence of generalized Darwinism, *Journal of Evolutionary Economics* 18: 577–596.
Baker, L.R. (2019): Just what is social ontology?, *Journal of Social Ontology* 5: 1–12.
Bidney, D. (1953): *Theoretical Anthropology*, New York: Columbia University Press.
Blute, M. (2010): *Darwinian Sociocultural Evolution: Solutions to Dilemmas in Cultural and Social Theory*, Cambridge: Cambridge University Press.
Boomsma, J.J. (2023): *Domains and Major Transitions of Social Evolution*, Oxford: Oxford University Press.
Bowler, P.J. (1983): *The Eclipse of Darwinism: Anti-Darwinian Evolution Theories in the Decades Around 1900*, Baltimore: Johns Hopkins University Press.
Bowles, S. & Gintis, H. (2011): *A Cooperative Species: Human Reciprocity and Its Evolution*, Princeton & Oxford: Princeton University Press.
Boyd, R. & Richerson, P. (1985): *Culture and the Evolutionary Process*, Chicago: University of Chicago Press.
Boyd, R. & Richerson, P. (2005): *Not by Genes Alone: How Culture Transformed Human Evolution*, Chicago: University of Chicago Press.
Bunge, M. (1977): *Treatise on Basic Philosophy, Volume 3 – Ontology I: The Furniture of the World*, Dordrecht: D. Reidel.
Buskell, A. (2025): Cultural lineages and the ranking problem, *Analysis*, Online advance access.
Cavalli-Sforza, L.L. & Feldman, M.W. (1981): *Cultural Transmission and Evolution: A Quantitative Approach*, Princeton: Princeton University Press.
DiFrisco, J. & Gawne, R. (2025): Biological agency: A concept without a research program, *Journal of Evolutionary Biology* 38: 143–156.
Epstein, B. (2015): *The Ant Trap: Rebuilding the Foundations of the Social Sciences*, New York: Oxford University Press.
Epstein, B. (2025): *Social Ontology*, Cambridge: Cambridge University Press.
Fábregas-Tejeda, A., Baedke, J., Prieto, G.I. & Radick, G. (Eds) (2025): *The Riddle of Organismal Agency: New Historical and Philosophical Reflections*, London & New York: Routledge.
Guala, F. (2016): *Understanding Institutions: The Science and Philosophy of Living Together*, Princeton & Oxford: Princeton University press.

Henrich, J. (2016): *The Secret of Our Success: How Culture Is Driving Human Evolution, Domesticating Our Species, and Making Us Smarter*, Princeton & Oxford: Princeton University Press.
Hiscock, P. & Sterelny, K. (2025): Beginnings: Africa and beyond, in: Ueki, T., Summerhayes, G.R. & Hiscock, P. (Eds): *In the Footsteps of Our Ancestors: Following Homo Sapiens into Asia and Oceania*, London & New York: Routledge, pp. 1–12.
Hodgson, G.M. & Knudsen, T. (2006): Why we need a generalized Darwinism and why a generalized Darwinism is not enough, *Journal of Economic Behavior and Organization* 61: 1–19.
Krohs, U. & Kroes, P. (Eds) (2009): *Functions in Biological and Artificial Worlds: Comparative Philosophical Perspectives*, Cambridge, MA: MIT Press.
Lewens, T. (2015): *Cultural Evolution*, Oxford: Oxford University Press.
Lewens, T. (2024): *Cultural Selection*, Cambridge: Cambridge University Press.
List, C. & Pettit, P. (2013): *Group Agency: The Possibility, Design, and Status of Corporate Agents*, New York: Oxford University Press.
Maynard Smith, J. & Szathmáry, E. (1995): *The Major Transitions in Evolution*, Oxford: Oxford University Press.
McCaffree, K. (2022): *Cultural Evolution: The Empirical and Theoretical Landscape*, London & New York: Routledge.
Mesoudi, A. (2011): *Cultural Evolution: How Darwinian Theory Can Explain Human Culture and Synthesize the Social Sciences*, Chicago & London: University of Chicago Press.
Mesoudi, A., Whiten, A. & Laland, K.N. (2006): Towards a unified science of cultural evolution, *Behavioral and Brain Sciences* 29: 329–347.
Morin, O. (2016): *How Traditions Live and Die*, New York: Oxford University Press.
O'Brien, M.J., Lyman, R.L., Mesoudi, A. & VanPool, T.L. (2010): Cultural traits as units of analysis, *Philosophical Transactions of the Royal Society B* 365: 3797–3806.
O'Connor, C. (2019): *The Origins of Unfairness: Social Categories and Cultural Evolution*, New York: Oxford University Press.
Okasha, S. (2018): *Agents and Goals in Evolution*, Oxford: Oxford University Press.
Okasha, S. (2024): The concept of agency in biology: Motivations and meanings, *Biological Theory* 19: 6–10.
Potochnik, A. (2017): *Idealization and the Aims of Science*, Chicago & London: University of Chicago Press.
Quine, W.V. (1948): On what there is, *Review of Metaphysics* 2: 21–38.
Reydon, T.A.C. (2025): *The Scope of Evolutionary Thinking*, Cambridge: Cambridge University Press.
Rouse, J. (2023): *Social Practices as Biological Niche Construction*, Chicago & London: University of Chicago press.
Runciman, W.G. (2009): *The Theory of Cultural and Social Selection*, Cambridge: Cambridge University Press.
Schmid, H.B., Hindriks, F., Ikäheimo, H., Laitinen, A., Salice, A. & Schweikard, D.P. (2015): Editorial note, *Journal of Social Ontology* 1: v.
Slade-Caffarel, Y. (2022): Economics and social ontology in Cambridge, *Syzetesis* 9: 313–330.
Spencer, H. (1890): *First Principles* (*A System of Synthetic Philosophy*, Vol. 1), New York: D. Appleton and Company.

Spencer, H. (1891): *Essays: Scientific, Political, & Speculative* (Vol. 1), London: Williams and Norgate.
Steward, J.H. (1955): *Theory of Culture Change: The Methodology of Multilinear Evolution*, Urbana & Chicago: University of Illinois Press.
Steward, J.H. (1958): Problems of cultural evolution, *Evolution* 12: 206–210.
Stoelhorst, J.W. (2014): The future of evolutionary economics is a vision from the past, *Journal of Institutional Economics* 10: 665–682.
Tomasello, M. (2022): *The Evolution of Agency: Behavioral Organization from Lizards to Humans*, Cambridge, MA: MIT Press.
Tomlinson, G. (2018): *Culture and the Course of Human Evolution*, Chicago & London: Chicago University Press.
Tuomela, R. (2013): *Social Ontology: Collective Intentionality and Group Agents*, New York: Oxford University Press.
Valković, M. (2024): Ontological and methodological limitations of certain cultural evolution approaches, *Philosophy of the Social Sciences* 54: 279–301.
Valković, M. (2025): *Power and Cultural Evolution* [Manuscript submitted for publication].
Weinstein, D. (2024): Herbert Spencer, in: Zalta, E.N. & Nodelman, U. (Eds): *The Stanford Encyclopedia of Philosophy* (Summer 2024 Edition), https://plato.stanford.edu/archives/sum2024/entries/spencer/.
White, L.A. (1943): Energy and the evolution of culture, *American Anthropologist* 45: 335–356.
White, L.A. (1959): *The Evolution of Culture: The Development of Civilization to the Fall of Rome*, New York: McGraw-Hill Book Company, Inc.

Part I

Metaperspectives

2 Evolutionary Theory in the Sociocultural Domain and Its Criticism

Simon Lohse and Martina Valković[1]

2.1 Introduction

There has long been a conviction that the theory of evolution, as developed by Darwin, is not limited to the biological realm, but that it can also be utilized elsewhere, including in such diverse spheres as science and technology, medicine, archaeology, and physics (Du Crest et al., 2023). In particular, evolutionary approaches to socio-cultural phenomena, traditionally investigated by the social sciences and the humanities, abound. There are evolutionary approaches aiming to understand the emergence of specific types of social behavior, such as cooperation (Henrich & Henrich, 2006; Boyd & Richerson, 2009; Tomasello et al., 2012; Bowles & Gintis, 2013; Molleman et al., 2013; Richerson et al., 2016), as well as xenophobia or privileging in-group members (Faulkner et al., 2004; Hammond & Axelrod, 2006; Efferson et al., 2008; Schaller et al., 2010; Fu et al., 2012; McCaffree, 2022; Sunsay, 2023), or imitating "prestigious" people (Henrich & Gil-White, 2001; Maner & Case, 2016; Brand & Mesoudi, 2019; Jiménez & Mesoudi, 2021). Evolutionary approaches are also used to predict or explain the change in social institutions such as marriage and family ties (Fortunato, 2015; Henrich, 2016) or institutions of sharing (Henrich, 2016; Guttentag, 2023; Andrews et al., 2024), to explore the transmission of cultural products such as material artefacts and literature (Ziman, 1999; O'Brien et al., 2001; Prentiss, 2011; Nikolić, 2018; Maier et al., 2023), and there are even attempts to explain the development of whole societies (Turchin, 2018).

These very diverse research programs have something in common. All of them use concepts, research strategies, and/or methods imported – or borrowed – from evolutionary theory in biology. These concepts, strategies, and methods are then adapted and used to understand certain aspects of socio-cultural reality, in particular dissemination dynamics (e.g., of behavioral traits or ideas) and socio-cultural diversity, inertia, and change. One of the main motivations for this transfer of evolutionary thinking is

DOI: 10.4324/9781003643845-3

that it is believed to provide a better basis for understanding socio-cultural phenomena and processes, by complementing, grounding, or even *replacing* traditional approaches in the social and human sciences. Some of the purported advantages of the evolutionary approaches over those of more traditional social sciences are their reliance on formal modeling and emphasis on quantitative methods, which is claimed to result in a rigorous and "fully scientific" method, as well as a common theoretical framework with the potential of synthesizing the social sciences (Mesoudi, 2011: 18–23; for a systematic criticism of these claims, see Lohse, 2023). Transfer of evolutionary thinking from biological to social sciences is at times motivated by the belief that social sciences are, essentially, biological sciences, justified by the fact that humans are biological beings (Rosenberg, 2017), or by the belief that social sciences have stagnated while biology has progressed, which is taken to suggest that theory transfer from biology will help to advance the social sciences (Mesoudi et al., 2006).

Evolutionary approaches to socio-cultural reality (in short "EvoSoc") have not only found supporters, but have often been viewed with skepticism and a critical eye, too. Critics have attacked evolutionary approaches from an ontological, epistemological, or political angle, highlighting different ways in which EvoSoc can be unhelpful, misleading, and even downright problematic. It has been doubted, for example, whether there are sufficient ontological parallels in the natural and social spheres to justify the transfer of evolutionary thinking (Schatzki, 2001; Reydon & Scholz, 2009, 2015). And evolutionary programs have been criticized for various explanatory deficits and for having problematic implications for society (Fracchia & Lewontin, 1999, 2005; Ingold, 2007; Cofnas, 2018; Chellappoo, 2021a; Lohse, 2023; Valković, 2023).

In this chapter, we want to contribute to this critical discourse. We will not, however, add a new type of argument to the list of criticisms or elaborate on a specific issue (see Nichols et al., 2024, for an overview of conceptual, methodological, and theoretical issues in EvoSoc). Instead, we aim to make a clarifying contribution to the discussion by mapping and assessing critical arguments and their connections. We focus on arguments that target typical assumptions and broader tendencies of EvoSoc, often from different angles. Our aim is to better understand the problematic aspects of these criticized assumptions and tendencies and to identify interrelations between different angles (or types) of critique. The main motivation for this aim is the conviction that a more thorough mapping of the discursive landscape will lead to a better understanding of the prospects and limitations of the evolutionary approaches in question. That is, a more detailed and systematic discussion and assessment of criticisms aimed at EvoSoc will enable us to assess more clearly the potential of these approaches to explain socio-cultural phenomena, as well as the potential pitfalls they bring. It will also help to understand why, on the whole, evolutionary research programs have not been able to become mainstream paradigms in social science and humanities fields (Du Crest et al., 2023). To achieve these goals, we proceed as follows. In Section 2.2, we explain and situate our approach in more detail. Section 2.3 is devoted to mapping and assessing arguments that take a critical look at common assumptions of EvoSoc. We focus on assumptions regarding individualism, agency, and natural selection of institutional entities. Section 2.4 concludes with a summary of key points and a reflection on the implications of our discussion.

2.2 Approach

In this section, we want to explain our approach in more detail and relate it to other critical contributions to EvoSoc. Critical contributions in our field of interest are often either all-encompassing and fundamentally reject EvoSoc or – just the opposite – very detailed, focusing on specific details of a certain approach (see, however, Chellappoo, 2021b; Cofnas, 2018). Consider this quote by Fracchia and Lewontin:

> Cultural evolutionary theories are carefully constructed, logically consistent, and very neat. Their neatness, however, is achieved either by dismissing as inessential to cultural evolution the contingencies that are so essential to historical change or by subsuming them to a single transhistorical principle of explanation. But this formulaic treatment is *fully inappropriate* to the labyrinthine pathways, the contingent complexity, the many nuances, and general messiness of history.
>
> (Fracchia & Lewontin, 1999: 77–78, emphasis added)

This is an example of an extremely comprehensive criticism of the application of evolutionary thinking in social and cultural science fields, claiming its inappropriateness on epistemological grounds. Or, consider the following criticism of Mesoudi et al. (2006) on both theoretical and political grounds:

> The genealogy of cultural traits that Mesoudi et al. propose . . . is *a parody of history* in which agency, power and social relations are but the ephemeral effects of proximate causes whose ultimate source is supposed to lie in capacities and dispositions bequeathed to individuals as an ancestral legacy, independently and in advance of their life in the world. It is ironic that while Mesoudi et al. are quick to dismiss the "erroneous" view . . . that cultural evolution entails a progressive advance from primitive beginnings to modern science and civilization, precisely such an advance is presupposed by the distinction, *upon which their*

> *entire scientific project is based*, between people in "traditional" communities whose behavior is governed by evolved traits, and rational people like themselves who are in a position to study them. Rejecting the idea of progress as a measure of cultural evolution does not prevent them . . . from unreservedly endorsing the idea when it comes to the history of their own science. They appear to be blind to the political implications of such double standards.
>
> (Ingold, 2007: 14, emphasis added)

In this instance, a cultural evolutionary approach to the study of culture is wholly dismissed as both inadequate for its stated purpose and politically problematic.

In contrast to these fairly broad-brush criticisms, there are also more focused criticisms centered on specific details or tools of certain evolutionary approaches. For instance, Lewens (2023) argues that the Price equation, a general analytical tool widely used and appreciated within evolutionary theory, can lead to various misleading results when applied in the domain of culture, and suggests limiting its use to cases in which selection and transmission can be fairly easily distinguished. This is an example of a specific, limited criticism intended to show that a specific tool or method does not have general applicability across cultural evolutionary contexts. Another example of this kind is that of Rubin (2023), who similarly cautions on limitations of borrowing models from biology into economics due to the differing measures of correlation in the two fields, which can lead to misunderstandings, unreliable heuristics, and misuse. Both of these criticisms target a rather small group of cultural evolutionary approaches, those that utilize, respectively, the Price equation and measures of correlation. These methods and models are by no means typical or widespread in EvoSoc, making these criticisms, accordingly, rather specific and not relevant for the majority of EvoSoc.

We want to do something different in this chapter, which is to offer an analysis situated between these two poles. In other words, we want to develop a *mid-range perspective* that assesses and connects arguments that target core assumptions and broader tendencies that are widespread, although not ubiquitous, in EvoSoc. We are thus concerned neither with arguments targeting only single aspects of approach X nor with those targeting the overall endeavor of transferring evolutionary thinking to the sphere of the humanities and social sciences. Rather, we want to zoom in on arguments that criticize *typical* assumptions that are present, in various forms, in different research programs. (We come back to this point in the concluding section.) We will make clear to what extent these typical assumptions can be criticized in various respects and also how these respects are connected. In addition, we will highlight key presuppositions that the specific arguments have to make in order to highlight potential limitations and difficulties these arguments face.

Our focus is on evolutionary approaches that can be considered well-established and that are in direct competition with established research programs in the humanities and social sciences. Under "well-established" we mean, primarily, that a certain consensus has formed around core assumptions of the respective approaches and that they have succeeded in gaining proponents and advancing a research agenda (Lakatos, 1978). That they are in "direct competition" with established research programs in the humanities and social science is not meant to imply that the respective representatives of these approaches necessarily compete with each other (indeed, there is more often a lack of engagement rather than competition), but rather that they propose alternative and often incompatible methods and assumptions when it comes to studying the phenomena of interest. Here, we include approaches in the fields of evolutionary psychology, institutional evolution theory, organization ecology, and various brands of dual inheritance theory and their offspring (Mesoudi, 2011). We will, on the other hand, not discuss evolutionary approaches that make very moderate or generic claims, that could be summed up as "human behavior is to some extent rooted in evolution", or that use evolutionary talk only metaphorically without making serious explanatory claims to the effect that culture *actually* evolves. Rather, we engage with positions that propose serious explanatory alternatives to traditional approaches in the social sciences and humanities. This focus on typical assumptions of well-established and ambitious evolutionary research programs helps to avoid cherry-picking and raising strawmen.

As noted earlier, the various criticisms targeted at EvoSoc can be divided into ontological, epistemological, and political variants. By ontological criticisms, we mean those critical arguments that target entities or processes postulated or assumed by cultural evolutionary theories, as well as their properties. One could criticize, for example, that postulated units of evolution do not do justice to specific features of the social world, such as its meaning dimension. By epistemological criticisms, we have in mind arguments that question the viability or usefulness of cultural evolutionary explanations or the knowledge generated by them. Here, we also include cases when certain alternative explanations are precluded by the typical assumptions or tendencies of EvoSoc. For instance, the usefulness of cultural evolutionary approaches for explaining the spread of innovations that do not take into account the role of institutional, regulatory, or profit-seeking factors may be considered limited. Under political criticisms, we include arguments that problematize political and societal assumptions or implications of cultural evolutionary thinking in the sociocultural domain, as well as arguments targeting policy suggestions made by their proponents or implicit in the approach. If one criticizes evolutionary psychology for promoting sexist stereotypes, that is political.

The division into these three categories is not intended to be taken as if the critical arguments can always be strictly and neatly separated or as if the same criticism cannot have an ontological, epistemological, and political aspect. Indeed, one of our aims is to show how these various criticisms are connected and interrelated; how these aspects interact and overlap. The value of this threefold, heuristic categorization is that it clarifies the different kinds of grounds cultural evolutionary approaches can and have been criticized on.

2.3 Key Assumptions in Evolutionary Approaches to Sociocultural Reality

In this section, we introduce assumptions and broader tendencies that are present in several EvoSocs, and discuss types of arguments that address these from at least two of the mentioned ontological, epistemological, and political perspectives. We also emphasize connections between the arguments. First, we map and assess criticism of reductive individualism, then of prevalent human agency concepts in EvoSoc, and finally, we look at arguments targeting the concepts of natural selection of institutional entities.

2.3.1 Reductive Individualism

EvoSoc often relies on a research strategy that maintains "that complex socioecological phenomena are fruitfully studied piece by piece – in a reductionist rather than holistic fashion" (Alden Smith, 2000: 29). This strategy often amounts to a form of reductive individualism. Here is a quote from Lewens summarizing the key idea:

> Cultural evolutionists are concerned with giving analyses of broad social institutions. . . . Their analyses are explicitly reductionist, in the sense that they aim to show how the stability of institutional practices can result from the aggregated behaviors of individuals.
>
> (Lewens, 2015: 139).

Although not all EvoSoc share this assumption, it can be clearly recognized in dual inheritance theories and evolutionary psychology (Richerson & Boyd, 2005; Mesoudi, 2011; Tooby & Cosmides, 1989; Cosmides & Tooby, 1989; Richerson & Henrich, 2012; Henrich, 2016). The idea is to provide individualistic explanations for the emergence, persistence, and/or change of socio-cultural phenomena such as social norms and institutions. These phenomena are to be explained through an analysis of the interactions between individuals and with the help of the toolbox that evolutionary thinking provides in order to provide proper microfoundations for the social sciences (Richerson & Henrich, 2012). For instance, Richerson and Boyd claim in their seminal book *Not by Genes Alone* that cultural change comes about through the everyday acquisition and utilization of cultural information by individual actors and that an explanation of socio-cultural macro-processes at the level of populations will need to ground these phenomena in an account describing interacting individuals (cf. Richerson & Boyd, 2005: 6, 123, 246). Mesoudi (2011: 51) argues that this approach can bridge the so-called micro–macro divide in the social sciences. According to Mesoudi, there has been a tendency either to study the psychology of individuals in isolation, without paying attention to the ways in which culture influences their behavior, or, just the opposite, to study cultural macro-processes and patterns without paying attention to the individual interactions giving rise to these. His version of an EvoSoc aims to change this by using evolutionary tools to explain socio-cultural phenomena through the social interaction between individuals. Paradigmatic for this approach are "transmission chain experiments", which aim to explain cultural patterns and their effects on actors through individualistic interaction models and specific learning biases (Mesoudi & Whiten, 2008).

In a similar vein, evolutionary psychologists aim to ground explanations of social institutions and cultural macro-patterns in models of interacting individuals in different environments. A central motive for this is their mistrust in the explanatory power of macro-level processes and forces, which, according to Cosmides and Tooby, remain mysterious if not anchored in accounts of individual representations and behavior. The distinctive feature of their approach is the idea that the behavior of individual actors and their social dynamics can ultimately be traced back to certain mental mechanisms that are adaptations shaped by evolutionary processes (Cosmides & Tooby, 1989, 1994; see below for more details).

This is textbook reductive individualism as it has been defended in the philosophy of the social sciences, often under the name of methodological individualism. Note that this brand of individualism combines two claims: one ontological and one methodological. It posits that socio-cultural macro-phenomena and their causal effects are *realized by* (nothing but) individuals interacting, exchanging ideas and goods, etc. (ontological claim), and that *therefore* we should aim for *reductive explanations* tracing this ontological dependency relation (methodological claim). Tooby & Cosmides (1992: 47) make this connection most explicit when criticizing what they call the "Standard Social Science Model" for abstracting away from a proper ontological micro-foundation and hence from a mechanism that could actually explain.[2]

There are at least three types of criticism levelled at the outlined individualistic tendency in EvoSoc. *First*, some authors criticize the approaches for neglecting essential aspects of socio-cultural reality, thus creating a distorted ontology. Two points in particular are frequently cited in this context.

(a) EvoSoc are said to ignore the "holistic meaning" dimension of culture (Ingold, 2007: 16; Fracchia & Lewontin, 1999, 2005). This is often expressed as the idea that cultural elements cannot really be understood in isolation – as cultural information bits that can be learned from each other in isolation – but often can only be understood as interwoven in a cultural web of meaning. Of course, there are cases where a cultural transfer of certain elements is possible through social learning (it is these cases that are often used to illustrate EvoSoc). Think of how you can learn a new way of fishing through observing successful role models. But there are many cases where this does not seem to work. For example, you cannot learn in the same way to submit a vote of no confidence to dismiss a government. This process needs to be understood as an interconnected web of actions, including speech acts, and concepts that have meaning only in a specific and actively designed institutional and legal context. The clue to this argument lies in the claim that at least certain cultural elements cannot be treated in isolation from one another in the way that individualistic EvoSoc does. This point is thus directed against the reductionist idea that complex cultural phenomena are just simple aggregates of individual cultural elements. According to the holistic argument, this ontology misses the holistic character of culture, which has to be understood as a web of interconnected, meaningful, conceptual threads (Geertz, 1973; cf. Mandelbaum, 1955; Zahle, 2016).

(b) Aiming to explain the workings of the social world by focusing solely on individuals, their attributes, and aggregated effects of their actions also overlooks, it has been argued, the existence and effects of social structures, and can therefore be characterized as a distortion of social reality (Lewens, 2015: ch.7; Cofnas, 2018; Fracchia & Lewontin, 1999, 2005; see also Chapter 4 of this volume, Section 4.5). In contrast to the EvoSoc image, individuals are here seen as born into a world of institutions, social systems, and economically and legally structured social relations which enable and constrict them, through a variety of systematic effects (Lukes, 2005). Social structures shape the perception and preferences of individuals (Lukes, 2005), channel the basic ways in which they interact with each other (Wartenberg, 1990), and may even play an important role in the constitution of individuals as agents (Lukes, 2005; Wartenberg, 1990). Reconsider the example of a no-confidence motion. This action may not only be semantically dependent on a certain social context, but it is made *possible* only by its embedding in legal and political structures (including legal texts and procedural guidelines). In the literature, in classical sociological theory, this idea has been described as a kind of intertwinedness of structure and action (Bourdieu, 1977; Giddens, 1984). Other authors go even further and assume that social structures can have autonomous causal effects on individual actors by way of emergent downward causation (see Sawyer, 2005: chapter 5 for an overview). Both (a) and (b) amount to attesting EvoSoc as an under-complex conception of social reality that cannot do justice to its more holistic or structural aspects.

A *second* type of argument targets the epistemological dimension of individualistic EvoSoc by criticizing it for providing misleading or spurious explanations. Take, for example, a cultural evolutionary explanation for the spread of hybrid corn in the United States, according to which it is due to individuals copying each other's attractive behaviors (Henrich, 2001; Mesoudi, 2011). An alternative explanation of this phenomenon has been proposed by Richard Lewontin (1992), who gives a central role to commercial interests of hybrid seed producers and traders, as well as governmental structures and policies. The key idea here is that an institutional framework was put in place that structured the space of adaptive possibilities for farmers and thus also social learning patterns. An explanation along these lines is simply not available from this kind of cultural evolutionary perspective, as it de-emphasizes or disregards the role of organized agents, institutional factors, and the like, in social and cultural change. According to this argument, individualistic EvoSoc explanations may therefore blind us with respect to the role and importance of these factors for rich explanations of social change.

The *third* type of argument targets individualistic EvoSoc from a political angle. This is perhaps most striking in cases in which EvoSoc assert or imply that social inequality is (primarily, mostly, predominantly, and so on) a result of individual choices, influenced by evolved psychology. For instance, in some recent EvoSoc work (Browne, 2023; Stern & Madison, 2022), it is claimed that occupational choice, as well as workplace disparity between women and men (including phenomena such as the pay gap, glass ceiling, and under-representation of women in STEM), is due to sex differences in evolved psychology. The explicitly stated aim of these authors is to influence policy by making claims about people's temperament, abilities, and interests based on an EvoSoc theory of the evolution of their sex. In addition to what appears to be a less than solid empirical basis for these claims – and the minimization of the potential influence of social, familial, institutional, and other factors in the development of temperament, interests, and abilities, as well as in enabling or constraining choices and opportunities (see, e.g., Dupré, 2001: 56–59 for a systematic criticism of evolutionary psychology in these respects) – this approach could also be criticized for committing an is-ought fallacy. Even if we do grant, for the sake of the argument, that there are evolved sex differences in the above-listed domains, this still does not imply that these differences *should* be respected, enabled, or encouraged, rather than discouraged or reined in (compare this case with the treatment of evolved preferences for eating foods rich in sugar and fat).

These arguments do indeed point to different ways in which individualistically oriented EvoSoc can be problematic. They may paint an undercomplex picture of social reality (ontology), ignore central explanatory factors (epistemology), or be problematic for political reasons. At the same time, it seems straightforward to identify connections in the arguments outlined. What exactly is the reason why it can be problematic to ignore social-structural aspects in EvoSoc explanations? Here, of course, it can be highlighted that this is problematic precisely because it fails to capture an essential factor of social reality – a factor that actually has a massive influence on the behavior of individual actors. The alleged problem here lies in the fact that the explanation does not reflect the actual (causal) relationships in social reality. An analogous connection can be drawn to the sketched political criticism.

Interestingly, however, the epistemological argument can also be seen as independent of the ontological argument, namely insofar as it does not necessarily have to be based on explanatory realism (i.e., the idea that explanations have to map true ontological dependency relations), but can be developed in line with instrumental pragmatism. One can, for example, question the explanatory power of individualistic EvoSoc by trying to show that their explanations are not as fruitfully applicable to other situations as structuralist explanations or by trying to show that explanations with incompatible (e.g. social structuralist) ontologies do a better job given the same – or a different – epistemic goal (see Danks, 2015, for a defense of ontological pluralism along the latter lines). This move would make the epistemological critique independent of the ontological critique, which brings with it an argumentative advantage. The reason for this is that the outlined ontological critique needs to make certain (contestable) presuppositions. The first is that the criticism put forward is based on a more adequate conception of social reality. However, as the decades-long debate about individualism/holism in the social sciences shows (see, e.g., Kincaid, 1986, 1993; Van Bouwel & Weber, 2008a, 2008b; Zahle & Collin, 2014; Lohse, 2017; Lauer, 2019), this is far from an innocent premise. The second presupposition is that EvoSoc is actually concerned with developing an adequate social ontology. Some of its proponents might dispute this and qualify their project as a simplifying model to explain only certain social phenomena.[3] Note that this gambit does not seem to work for the political argument. For, in this case, the normative claims are based on the assumption that the postulated evolutionarily developed preferences are indeed mapping how things really are.

2.3.2 *Human Agency*

Another common tendency in EvoSoc is to treat human agency in a reductionist way or to de-emphasize its importance, typically by shifting focus to some other factors internal or external to the individual. In the provided image of human beings, cultures and societies, human agency, understood as the ability to make a difference in the world through intentional and (at least to some degree) autonomous action, does not feature a prominent role.

Approaches in which this kind of image is perhaps the most evident include E.O. Wilson's sociobiology, evolutionary psychology, and dual-inheritance theory. We focus here primarily on the last two, more recent programs. Evolutionary psychologists (Barkow et al., 1992; Buss, 2014; Cosmides & Tooby, 1989, 1994) focus on the evolutionary history of our species, rather than on individual agency, when it comes to explaining behavior. In their view, our behavior is a result of our evolved modular psychology, basically a large collection of specialized cognitive mechanisms that are adaptations resulting from the evolutionary history of our species. The agency in this account is mainly attributed to natural selection, which shaped these cognitive mechanisms (see earlier). Ariew & Panchanathan (2023: 495) write that in evolutionary psychology, "a species-typical genotype is exposed to a set of environmental conditions which results in an unfolding process of development", a process in which organismal agency does not feature as a particularly relevant factor.

Dual-inheritance theories (Richerson & Boyd, 2005; Mesoudi, 2011; Richerson & Henrich, 2012; Henrich, 2016) give central importance to various internal and external forces that also seem to preclude the idea of humans as autonomous agents. On the internal side, there are various subpersonal-level biases and tendencies, such as prestige bias and conformist bias, which are typically considered to be genetically evolved. These are taken to influence the way in which individuals acquire cultural traits, which is largely thought to happen via social learning (which is also considered to be an innate tendency), understood as quasi-automatic copying of the behavior observed in others. This is where the external forces come into play, by means of a cultural environment constituted by observable behavior and artefacts, which are decisive when it comes to which cultural traits the individual will copy and acquire. In this manner, the internal and external combine to shape the individual action.

In both research programs, therefore, agency is not considered as the human capacity for intentional and (partially) autonomous action. Rather, in evolutionary psychology, agency is attributed to natural selection, and in the dual-inheritance account, to both natural and cultural selection.

As with the individualistic tendencies described in the previous section, the tendency to focus on factors other than individual agency can also comprise both ontological and methodological claims. In the case of evolutionary psychology, the ontological claim is that our psychology just *is* modular, that is, that our brains are collections of evolved highly specialized adaptive cognitive mechanisms, and that therefore – this is the methodological conclusion – the best way to study human psychology and behavior is by modeling it as modular (combined with the individualism assumption, this results in the claim that focusing on these psychological modules is also the best way to study human societies). There is a parallel case in dual-inheritance theory.

We now come to various critical arguments against the described ideas on human agency. *First*, it has been criticized on ontological grounds that by neglecting agency, a central feature of human existence and our social lives is missing, and therefore a misleading ontology is put forward. One example of such criticism is Dupré (2001), who attacks evolutionary psychologists such as Buss, Tooby, and Cosmides for not paying attention to human autonomy. He claims that this follows from the assumptions that their theory shares with much of contemporary science and philosophy (Dupré, 2001: 17), such as privileging intrinsic properties of an organism instead of those that emerge from social interactions and context (Dupré, 2001: 72). As opposed to the evolutionary psychology image, Dupré makes a case for a stronger form of human autonomy in the sense of agent causation, where humans are sources of causality due to their ability to choose and follow principles that they develop on the basis of social interactions.

Another example of ontological critique is aimed at a broader field of research programs. Ariew & Panchanathan (2023) focus on not only all biologists but also social scientists who apply Niko Tinbergen's approach to the study of human agency (among whom they include sociobiologists (Wilson, 1975), evolutionary psychologists (Barkow et al., 1992; Buss, 2014; Pinker, 2003), and cultural evolutionists (Boyd & Richerson, 1985; Cavalli-Sforza & Feldman, 1981)). Ariew and Panchanathan claim that this approach is committed to treating biological organisms as objects instead of purposive agents, thereby missing an important feature of many organisms, namely, that they *are* purposive agents (Ariew & Panchanathan, 2023: 479). Ariew and Panchanathan claim that Darwinian evolution, especially in the post-Modern Synthesis era, operates with ontological commitments to treat individuals as material objects with intrinsic causal dispositions, including dispositions to behave in certain ways in certain environmental conditions (Ariew & Panchanathan, 2023: 487). As such, it represents an incomplete ground for studying (human and non-human) organismal behavior, since at least *some* organisms are genuine agents, both in the sense of being able to "actively participate in their own transformational processes and in modifying their environments" and in the sense of being conscious (Ariew & Panchanathan, 2023: 497).[4]

Second, the tendency to minimize agency has been criticized on epistemological grounds, too. Chellappoo (2021b, 2022) has addressed the epistemic dimension of agency minimization. Chellappoo (2022) targets the explanatory power of cultural selectionist approaches, in particular, with an emphasis on Henrich (2016). She makes a case for comparing cultural selection explanations to goal-directed agent explanations, claiming that the former will be useful tools only when the latter cannot already explain all relevant aspects of a certain phenomenon. This is so because, according to Chellappoo, the key component of a successful and powerful cultural selection explanation is the ability to explain "design without the designer", whereas, if the phenomenon can be explained by a goal-directed agent approach, it cannot anymore be designated as such, but rather as intentional "design". When cultural-selection and goal-directed agent approaches have the same explanatory power, the goal-directed agent approaches are to be preferred, since they are well established and have proven to be effective explanatory tools in a wide variety of contexts, and because humans *are* intelligent, goal-directed agents able to design and modify artefacts and behavior. Chellappoo (2021b) makes a similar case against prestige bias accounts, arguing that the prestige bias model, widespread in cultural evolutionary theories, does not represent an increase in explanatory power over a commonsense model of a goal-directed agent account that is both widespread and demonstrably effective in many domains. Chellappoo illustrates this with an example of a woman marrying into a new community and having to acquire a new skill of arrowhead making. On the prestige bias account, the woman would first determine who the most prestigious arrowhead maker in the community is, "paid the cost of access" in the sense of deference to the maker, and then copy her behavior. In the goal-directed agent model, by contrast, the woman could pursue different options available to her to learn this skill, such as consulting with friends, individual learning, and observation of successful makers. In this model, too, she could pay deference to the most prestigious maker, for instance, out of conscious calculation or genuine admiration. Chellappoo concludes that we should not switch to this less intuitive framework that downplays human agency since it may not feature increased explanatory or predictive power.

The *third* kind of criticism involves arguments from ethical and political angles. Ingold (2007) criticizes Mesoudi et al. (2006) and what he terms their neo-Darwinian paradigm on several accounts, including for being

> an affront to the millions of intelligent human beings for whom traditions are real and important but who are not, on that account, trait-bearing cultural clones whose only role in life is to express – in their

behavior, artefacts and organizations – information that has been transmitted to them from previous generations.

(Ingold, 2007: 14)

Ingold claims that Mesoudi et al.'s whole project is based on the distinction between people in "traditional" communities whose behavior is a consequence of evolved traits, and rational people like themselves who study them, with obvious (and familiar from our history) political implications of such double standards. Chellappoo (2021a) picks up on this criticism and develops it in terms of "othering" and exoticization, processes by which, by stripping human beings of their agency, there is a possibility to construct an "us" versus "them" distinction (i.e., the studier vs. the studied), as well as potential for dehumanization of people thereby designated as "them". The minimization of agency can then, she claims, negatively impact people whose cultural behavior EvoSoc aims to explain.

As we have shown, agency-minimizing EvoSoc are considered problematic in various ways, including by presenting a misleading ontology that omits or downplays an important feature of human beings, by putting forward agency-minimizing models that do not have more explanatory or predictive power compared to the more commonsense models of humans as goal-directed agents, and by having the potential for othering and dehumanization of the human beings studied. While these arguments can be endorsed (or opposed) independently from one another, there are noticeable connections between them, which are in some cases even explicitly stated. So, Ariew and Panchanathan claim that agency-minimizing EvoSoc are incomplete grounds for *studying* organismal behavior since (at least some) organisms *are* agents. Similarly, Chellappoo (2021b, 2022) argues that goal-directed agent approaches should be preferred to the agency-minimizing EvoSoc when they have the same explanatory power, since humans *are* goal-directed agents. In these cases, ontological concerns are taken to be relevant for epistemic projects. Of course, one could argue that goal-directed agents' approaches should be preferred over agency-minimizing ones on grounds other than ontological ones. In this context, different (mostly implicit) interpretations and weightings of epistemic values often play an important role, for example, when it comes to evaluating the respective scope of agency versus evolutionary explanations or balancing scope with intuitive intelligibility (see the epistemological arguments by Chellappoo).

Mirroring the discussion from the previous section, when it comes to contestable presuppositions of the above-presented critical arguments, there are again considerations of explanatory realism as opposed to instrumental pragmatism to be taken into account. Some might contest the claim that humans are actual, true agents, understood as being able to act in some sense independently from internal and external forces influencing them. Proponents of agency-minimizing EvoSoc might, as well, deny that they are in the business of providing adequate social ontology, and instead claim that they are just making a case for simplified models. While some EvoSoc might more plausibly make such claims, much of the dual-inheritance work, and especially evolutionary psychology, seems rather committed to stronger, ontological claims. The modules and other adaptations that supposedly shape our (social) behavior are considered to actually exist according to theories such as that of Tooby and Cosmides (e.g., 1989: 31; Cosmides & Tooby, 1989: 52) and Buss (e.g., 1994: 50–51), as do adaptations for social learning and group hierarchies in accounts such as those of Boyd and Richerson (e.g., Richerson & Boyd, 2005: 6), Henrich (e.g., 2016: 52, 142), or Mesoudi (e.g., 2011: 191). In the case of the political criticism to the effect that agency-minimizing EvoSoc can other or dehumanize members of populations or cultures it studies, one could object that EvoSoc does, or at least can, treat *any* human subject in the same way. For example, not only members of small, remote communities but also prominent scientists in big centers can be studied through the agency-minimizing EvoSoc lens, thereby collapsing the "us vs. them" construction between the studied and the studier – to this one could still object, of course, that agency-minimizing EvoSoc, in a sense, dehumanizes *all* human beings.

2.3.3 Natural Selection of Evolving Populations of Sociocultural Entities

Another approach popular in EvoSoc stands in contrast to the individualistic reductionism discussed earlier: population-based modeling of sociocultural entities. The core idea of this approach is to model the behavior of certain types of social entities, such as institutions and organizations, on a population level to provide explanations and make predictions about their emergence, persistence, and/or change.

In the field of evolutionary economics, for instance, Nelson and Winter (1982: 9) speak of populations of firms in which patterns of differential survival and growth can produce change in economic aggregates of these populations. Hodgson and Knudsen (2010: 28) describe populations of businesses, states, and other organizations and claim they exist as parts of complex systems in which they interact with each other and their environment, struggle to survive against scarce resources, and adapt and replicate.

Organizational ecologists attempt to understand why the existing diversity of *organizational types* exists, using a population-based approach. For instance, they try to explain why certain types of organizations survive better in a given institutional environment than others based on Darwinian principles (see the seminal work by Hannan & Freeman, 1977, 1989).

The general proposition is to treat firms, NGOs, and such (more precisely, their organizational form) as organism analogues forming populations that are subject to evolutionary pressure leading to Darwinian adaptation and selection processes:

> Taking our lead from . . . distinguished ecologists, we suggest that a population ecology of organizations must seek to understand the distributions of organizations across environmental conditions and the limitations on organizational structures in different environments, and more generally seek to answer the question, Why are there so many kinds of organizations?
>
> (Hannan & Freeman, 1977: 936)

Here, we focus primarily on organizational ecology as perhaps the most developed type of population-based approaches. As with the other tendencies we discussed, organization ecology has been criticized from an ontological and a political angle. Reydon and Scholz (2009, 2015) have developed a detailed critique of organizational ecology targeting the assumptions of ontological transfer that are implicit in this approach. For the explanations that are provided by this specific EvoSoc to work, they argue that organizational ecologists have to assume that populations of organizations can evolve analogous to populations of organisms. Reydon and Scholz then go on to demonstrate that organizations cannot fulfil this role because they do not reproduce in the right way. Rather, populations of organizations need to be considered as *sets* of organizations with similarities. While these similarities may spread via social learning or manager activities, this is not enough for a population-based Darwinian explanation to work. Rather, population-based explanations would have to deploy evolutionary mechanisms referring to the selective advantages of the actual *offspring* of organizations that form relatively isolated and cohesive populations. The upshot of the argument is that the ontological disanalogies between biological and sociological phenomena are too big (at least in this case) for the approach to succeed.

The political critique of organizational ecology points to possible risks when applying organizational ecology thinking to real-world practice (which, of course, is not an unlikely scenario given that organizational theory is an element in academic management education). Organizational ecology emphasizes the causal role of the selective pressure of the socio-economic and legal environments in structuring organizations and selecting the "fittest" organizations. The influence of intentional actions, on the other hand, is de-emphasized. This view could lead managers and other members of an organization to systematically underestimate their own influence on the development of the organization and to develop a "fatalistic" outlook (Morgan, 1986; Scholz & Reydon, 2008). In addition, it may encourage a moral assessment of thriving organizations that is tainted by an ideology of Social Darwinism (Morgan, 1986; Scholz & Reydon, 2008). The natural selection of the "fittest" organizations could be used to justify their behavior as legitimate, even if it is politically or ethically problematic, since (apparently) this very behavior has been selected by "natural law" as appropriate for our society. In a nutshell, the normative force of natural law is invoked here.

The arguments described both relate to issues of theory transfer. In the first case, it is about the unjustified transfer of the ontological assumptions of biological evolutionary theory to organizational research. In the second case, it is a problematic transfer of theory into practice. Are there further connections between these arguments? In one sense, the political arguments can be seen as independent from the ontological argument. Let us say the ontological criticism does not work. Organizational ecology has actually elaborated a functioning mechanism that can do the intended explanatory work and show that the environment has a significant influence on the structuring and survival of organizations. Nonetheless, such a view could lead to a situation where a manager's ability to act is ultimately downplayed more than would be appropriate. Scholz and Reydon, however, explicitly draw on a connection between ontological and political argument. It is precisely the influence of an unjustified transfer of evolutionary theory that leads to problems, since politically problematic decisions and assessments are made on the basis of misguided assumptions about social reality. Among other things, this could lead to a situation where managers not only slightly underestimate their influence on the trajectory of organizations but, based on erroneous reasons, fundamentally misconstrue their role as mere observers and communicators of organizational change shaped by natural selection. This connection to ontology increases the power of the political argument – but it is important to point out that the plausibility of the political argument ultimately depends on empirical evidence. It must be demonstrated that organizational theory does indeed have a significant influence on the moral evaluation of the behavior of organizations or the behavior of managers. The latter in particular could ultimately be guided much more by practical considerations than by theory, which would clearly limit the argument's power. The ontological argument also has limitations, namely that its primary target is approaches in organization ecology that aim at a true account of how organizational change, etc., actually works. To the extent that researchers may be merely interested in (and to some extent, able to) *predict* how organizational populations (in their terms) develop, the critique loses some of its force. Organizational research would not be the only scientific field in which predictions work on the basis of false assumptions.

2.4 Conclusion

In this chapter, our aim was to map and assess some of the critical arguments that have been raised against evolutionary approaches to culture and society, as well as the connections between them, in order to gain more clarity regarding the potential and limitations of these evolutionary approaches – and their criticism. We focused on arguments targeting three widespread tendencies or assumptions of EvoSoc, namely individualism, agency minimization, and natural selection of populations of institutional entities. We divided these critical arguments into ontological, epistemological, and political ones, while acknowledging their interrelations and (in)dependencies. So, in the case of the individualistic tendencies of EvoSoc, we pointed out that, while it can be independently objected that EvoSoc put forward an insufficiently complex image of social reality, disregards some of the central explanatory factors, or is politically problematic, there are straightforward connections between these arguments. One reason why the individualistic explanation, for instance, can be considered inadequate is that it may not reflect real causal connections. Similarly, in the case of agency-minimizing EvoSoc, we presented different angles of critique, such as that they may represent a misleading ontology which omits an important human capacity, that the models employed do not lead to an increase in explanatory or predictive power, and that certain approaches potentially dehumanize their subjects of inquiry, if not human beings more generally. As in the case of individualism, we pointed out both the fact that these arguments can be held independently, and their noticeable (and often explicit) connections: one of the reasons why agency-minimizing models could be considered incomplete or inadequate is that human beings *really are* agents. Finally, in the case of conceiving of institutional entities as appearing in evolving populations, we pointed to an ontological critical angle, namely that organizations do not reproduce in the way needed for them to constitute a Darwinian population. We also outlined political criticism to the effect that the influence of intentional actions can be systematically underestimated as a consequence of organizational ecology theory and that organizational success and failure can be naturalized in Social Darwinist tones. We noted not only that these criticisms can be held independently but also that a direct connection has been drawn between them, in that politically problematic assessments are being made on the basis of a flawed ontology.

In addition to the interrelations between the kinds of critical arguments raised against one of the three described tendencies, there are also noteworthy relations between the three tendencies themselves. Both the individualism present in EvoSoc and the agency minimization are examples of reductionism – the first reduces the social world to individuals and aggregate effects of their behavior, while the second reduces human behavior to a sum of internal and external pressures and their interactions. In this way, the first two tendencies we described both present a certain reductionist image of human beings and/or their societies. There is a connection to be drawn between the agency minimization and the image of evolving populations of organizational entities, too. The latter can be seen as yet another way of limiting human agency, in which it is the natural selection of organizations that holds the central role, and people are relegated to the role of more-or-less passive observers of institutional change. The nature of agency minimization is of a completely different kind in this case, though. Finally, there is an interesting point to be made about individualism and evolving organizational populations, in that they can be seen as opposed in a sense – the former puts focus on individuals and their behavior, and the latter on impersonal forces of natural selection that act on populations of organizations, apparently blending individuals into the background. This contrast makes it particularly clear that it is necessary for criticism of cultural evolution theory to pay attention to different types of tendencies present in EvoSoc, as associated challenges will differ, sometimes quite drastically, as well.

The kind of mapping we engaged in this chapter is important for several reasons. It makes apparent that evolutionary thinking lends itself to highlighting certain aspects of phenomena, ontologies, and types of explanations, while neglecting others, and shows how, in some cases, this can be limiting or problematic, often in more than one respect. This hints at a certain worldview that creeps in, or is perhaps even baked into, this kind of theorizing. Based on the reasoning from the previous paragraph, it appears that the issues raised in this chapter are not (completely) independent or isolated, but that there are common presuppositions or guiding ideals of reductionism and the minimization or neglect of human agency. If this is correct, our mapping endeavor brought to the surface key theoretical commitments of evolutionary approaches that *underlie* the tendencies we discussed, which is an important step in a thorough assessment of these approaches.

However, it is equally important to, once again, point out that different evolutionary programs will be problematic in different ways. Here, we largely focused on dual-inheritance theory and evolutionary psychology with respect to individualism and agency minimization, and organizational ecology with respect to evolving organizational populations. Reductionism and agency minimization appeared as two threads in this literature, but that does not mean this will necessarily generalize to other approaches labelled evolutionary. Additionally, as we discussed, criticisms will often be conditional on different background assumptions and epistemic goals. Not all theories will have as their goal accurately describing social reality or providing predictive models, for instance, and any attempt at appropriate criticism will have to take this into account. These are just some of the considerations which illustrate how a much more nuanced and complex debate is needed with respect to evolutionary thinking and its proper place in the humanities and social sciences. This chapter is meant to represent a step in that direction.

Notes

1 Both authors contributed equally to writing this chapter.
2 The close connection between ontological and methodological claims has frequently been criticized; see, for example, Zahle & Kincaid (2019).
3 It is not clear to what extent this objection works. There are many passages in programmatic EvoSoc writings (e.g., Tooby & Cosmides, 1989: 29–30; Cosmides & Tooby, 1989: 51–52; Richerson & Boyd, 2005: 247; Mesoudi, 2011: 3; Henrich, 2016: ch. 4) that suggest an *actual* ontological commitment in the sense of individualism.
4 It should be noted that this is a richer concept of agency than the one outlined above, which can be considered a sub-element of Ariew's and Panchanathan's concept.

References

Alden Smith, E. (2000): Three styles in the evolutionary analysis of human behavior, in: Cronk, L., Chagnon, N. & Irons, W. (Eds): *Adaptation and Human Behavior: An Anthropological Perspective*, New York: Aldine de Gruyter, pp. 27–46.

Andrews, J., Clark, M., Hillis, V. & Borgerhoff Mulder, M. (2024): The cultural evolution of collective property rights for sustainable resource governance, *Nature Sustainability* 7: 404–412.

Ariew, A. & Panchanathan, K. (2023): Adding agency to Tinbergen's four questions, in: Du Crest, A., Valković, M., Ariew, A., Desmond, H., Huneman, P. & Reydon, T.A.C. (Eds): *Evolutionary Thinking across Disciplines*, Cham: Springer, pp. 477–499.

Barkow, J.H., Cosmides, L. & Tooby, J. (Eds) (1992): *The Adapted Mind: Evolutionary Psychology and the Generation of Culture*, New York: Oxford University Press.

Bourdieu, P. (1977): *Outline of a Theory of Practice*, Cambridge: Cambridge University Press.

Bowles, S. & Gintis, H. (2013): *A Cooperative Species: Human Reciprocity and Its Evolution*, Princeton & Oxford: Princeton University Press.

Boyd, R. & Richerson, P.J. (1985): *Culture and the Evolutionary Process*, Chicago: University of Chicago Press.

Boyd, R. & Richerson, P.J. (2009): Culture and the evolution of human cooperation, *Philosophical Transactions of the Royal Society B: Biological Sciences* 364: 3281–3288.

Brand, C.O. & Mesoudi, A. (2019): Prestige and dominance-based hierarchies exist in naturally occurring human groups, but are unrelated to task-specific knowledge, *Royal Society Open Science* 6: 181621.

Browne, K.R. (2023): The disjunction between evolutionary psychology and sex-discrimination law and policy, *Evolution and Human Behavior* 44: 284–295.

Buss, D.M. (1994): *The Evolution of Desire: Strategies of Human Mating*, New York: Basic Books.
Buss, D.M. (2014): *Evolutionary Psychology: The New Science of the Mind*, Hove: Psychology Press.
Cavalli-Sforza, L. & Feldman, M. (1981): *Cultural Transmission and Evolution: A Quantitative Approach*, Princeton: Princeton University Press.
Chellappoo, A. (2021a): Cultural evolution: A case study in global epistemologies of science, in: Ludwig, D., Koskinen, I., Mncube, Z., Luana, P. & Reyes-Galindo, L. (Eds): *Global Epistemologies and Philosophies of Science*, London: Routledge, pp. 208–219.
Chellappoo, A. (2021b): Rethinking prestige bias, *Synthese* 198: 8191–8212.
Chellappoo, A. (2022): When can cultural selection explain adaptation?, *Biology & Philosophy* 37: 1–23.
Cofnas, N. (2018): Power in cultural evolution and the spread of prosocial norms, *The Quarterly Review of Biology* 93: 297–318.
Cosmides, L. & Tooby, J. (1989): Evolutionary psychology and the generation of culture, part ii: Case study: A computational theory of social exchange, *Ethology and Sociobiology* 10: 51–97.
Cosmides, L. & Tooby, J. (1994): Better than rational: Evolutionary psychology and the invisible hand, *American Economic Review* 84: 327–332.
Danks, D. (2015): Goal-dependence in (scientific) ontology, *Synthese* 192: 3601–3616.
Du Crest, A., Valković, M., Ariew, A., Desmond, H., Huneman, P. & Reydon, T.A.C. (2023): Generalizing Darwinism as a topic for multidisciplinary debate, in: du Crest, A., Valković, M., Ariew, A., Desmond, H., Huneman, P. & Reydon, T.A.C. (Eds): *Evolutionary Thinking across Disciplines*, Cham: Springer, pp. 1–17.
Dupré, J. (2001): *Human Nature and the Limits of Science*, Oxford: Clarendon Press.
Efferson, C., Lalive, R. & Fehr, E. (2008): The coevolution of cultural groups and ingroup favoritism, *Science* 321: 1844–1849.
Faulkner, J., Schaller, M., Park, J.H. & Duncan, L.A. (2004): Evolved disease-avoidance mechanisms and contemporary xenophobic attitudes, *Group Processes & Intergroup Relations* 7: 333–353.
Fortunato, L. (2015): Marriage systems, evolution of, in: Wright, J.D. (Ed): *International Encyclopedia of the Social & Behavioral Sciences*, Amsterdam: Elsevier, pp. 611–619.
Fracchia, J. & Lewontin, R.C. (1999): Does culture evolve?, *History and Theory* 38: 52–78.
Fracchia, J. & Lewontin, R.C. (2005): The price of metaphor, *History and Theory* 44: 14–29.
Fu, F., Tarnita, C.E., Christakis, N.A., Wang, L., Rand, D.G. & Nowak, M.A. (2012): Evolution of in-group favoritism, *Scientific Reports* 2: 460.
Geertz, C. (1973): *The Interpretation of Cultures: Selected Essays*, New York: Basic Books.
Giddens, A. (1984): *The Constitution of Society: Outline of the Theory of Structuration*, Oxford: Polity Press.
Guttentag, M.D. (2023): Evolutionary psychology and resource-sharing laws, *Evolution and Human Behavior* 44: 264–271.
Hammond, R.A. & Axelrod, R. (2006): The evolution of ethnocentrism, *Journal of Conflict Resolution* 50: 926–936.

Hannan, M.T. & Freeman, J. (1977): The population ecology of organizations, *American Journal of Sociology* 82(5): 929–964.
Hannan, M.T. & Freeman, J. (1989): *Organizational Ecology*, Harvard University Press.
Henrich, J.P. (2001): Cultural transmission and the diffusion of innovations: Adoption dynamics indicate that biased cultural transmission is the predominate force in behavioral change, *American Anthropologist* 103: 992–1013.
Henrich, J.P. (2016): *The Secret of Our Success: How Culture Is Driving Human Evolution, Domesticating Our Species, and Making Us Smarter*, Princeton & Oxford: Princeton University Press.
Henrich, J.P. & Gil-White, F.J. (2001): The evolution of prestige: Freely conferred deference as a mechanism for enhancing the benefits of cultural transmission, *Evolution and Human Behavior* 22: 165–196.
Henrich, J.P. & Henrich, N. (2006): Culture, evolution and the puzzle of human cooperation, *Cognitive Systems Research* 7: 220–245.
Hodgson, G.M. & Knudsen, T. (2010): *Darwin's Conjecture: The Search for General Principles of Social and Economic Evolution*, Chicago: University of Chicago Press.
Ingold, T. (2007): The trouble with 'evolutionary biology', *Anthropology Today* 23: 13–17.
Jiménez, Á.V. & Mesoudi, A. (2021): The cultural transmission of prestige and dominance social rank cues: An experimental simulation, *Evolutionary Psychological Science* 7: 189–199.
Kincaid, H. (1986): Reduction, explanation, and individualism, *Philosophy of Science* 53: 492–513.
Kincaid, H. (1993): The empirical nature of the individualism-holism dispute, *Synthese* 97: 229–247.
Lakatos, I. (1978): *The Methodology of Scientific Research Programmes*, Cambridge University Press.
Lauer, R. (2019): Is social ontology prior to social scientific methodology?, *Philosophy of the Social Sciences* 49: 171–189.
Lewens, T. (2015): *Cultural Evolution: Conceptual Challenges*, Oxford: Oxford University Press.
Lewens, T. (2023): Equations at an exhibition: On the cultural Price Equation, in: du Crest, A., Valković, M., Ariew, A., Desmond, H., Huneman, P. & Reydon, T.A.C. (Eds): *Evolutionary Thinking across Disciplines*, Cham: Springer, pp. 353–373.
Lewontin, R.C. (1992): *Biology as Ideology: The Doctrine of DNA*, New York: Harper Perennial.
Lohse, S. (2017): Pragmatism, ontology, and philosophy of the social sciences in practice, *Philosophy of the Social Sciences* 47: 3–27.
Lohse, S. (2023): Pluralism and epistemic goals: Why the social sciences will (probably) not be synthesised by evolutionary theory, in: du Crest, A., Valković, M., Ariew, A., Desmond, H., Huneman, P. & Reydon, T.A.C. (Eds): *Evolutionary Thinking across Disciplines*, Cham: Springer, pp. 333–351.
Lukes, S. (2005): *Power: A Radical View* (Second Edition), London: Palgrave Macmillan.
Maier, A., John, R., Linsel, F., Roth, G., Antl-Weiser, W., Bauer, L., Buchinger, N., Cavak, L., Hoffmann, H., Puschmann, J., Schemmel, M., Schmid, V.C., Simon, U. & Thomas, R. (2023): Analyzing trends in material culture evolution – a case study of Gravettian points from lower Austria and Moravia, *Journal of Paleolithic Archaeology* 6: 15.

Mandelbaum, M. (1955): Societal facts, *The British Journal of Sociology* 6: 305–317.
Maner, J.K. & Case, C.R. (2016): Dominance and prestige, in: Olson, J.M. & Zanna, M.P. (Eds): *Advances in Experimental Social Psychology* (Vol. 54), Cambridge, MA & San Diego, CA: Elsevier, pp. 129–180.
McCaffree, K. (2022): *Cultural Evolution: The Empirical and Theoretical Landscape*, New York & London: Routledge.
Mesoudi, A. (2011): *Cultural Evolution: How Darwinian Theory Can Explain Human Culture and Synthesize the Social Sciences*, Chicago: University of Chicago Press.
Mesoudi, A. & Whiten, A. (2008): The multiple roles of cultural transmission experiments in understanding human cultural evolution, *Philosophical Transactions of the Royal Society B: Biological Sciences* 363: 3489–3501.
Mesoudi, A., Whiten, A. & Laland, K.N. (2006): Towards a unified science of cultural evolution, *Behavioral and Brain Sciences* 29(4): 329–347.
Molleman, L., Quiñones, A.E. & Weissing, F.J. (2013): Cultural evolution of cooperation: The interplay between forms of social learning and group selection, *Evolution and Human Behavior* 34: 342–349.
Morgan, G. (1986): *Images of Organization*, Thousand Oaks: Sage Publications.
Nelson, R.R. & Winter, S.G. (1982): *An Evolutionary Theory of Economic Change*, Cambridge: Harvard University Press.
Nichols, R., Charbonneau, M., Chellappoo, A., Davis, T., Haidle, M., Kimbrough, E.O., Moll, H., Moore, R., Scott-Phillips, T., Purzycki, B.G. & Segovia-Martin, J. (2024): Cultural evolution: A review of theoretical challenges, *Evolutionary Human Sciences* 6: e12.
Nikolić, A. (2018): Artefacts in cultural transmission: Evolutionary perspective, *Etnoantropološki Problemi/Issues in Ethnology and Anthropology* 13: 1165–1182.
O'Brien, M.J., Darwent, J. & Lyman, R.L. (2001): Cladistics is useful for reconstructing archaeological phylogenies: Palaeoindian points from the Southeastern United States, *Journal of Archaeological Science* 28: 1115–1136.
Pinker, S. (2003): *The Blank Slate: The Modern Denial of Human Nature*, London: Penguin Books.
Prentiss, A.M. (2011): Introduction to this special issue on the evolution of material culture, *Evolution: Education and Outreach* 4: 374–378.
Reydon, T.A.C. & Scholz, M. (2009): Why organizational ecology is not a Darwinian research program, *Philosophy of the Social Sciences* 39: 408–439.
Reydon, T.A.C. & Scholz, M. (2015): Searching for Darwinism in generalized Darwinism, *The British Journal for the Philosophy of Science* 66: 561–589.
Richerson, P.J., Baldini, R., Bell, A.V., Demps, K., Frost, K., Hillis, V., Mathew, S., Newton, E.K., Naar, N., Newson, L., Ross, C., Smaldino, P.E., Waring, T.M. & Zefferman, M. (2016): Cultural group selection plays an essential role in explaining human cooperation: A sketch of the evidence, *Behavioral and Brain Sciences* 39: e30.
Richerson, P.J. & Boyd, R. (2005): *Not by Genes Alone: How Culture Transformed Human Evolution*, Chicago: University of Chicago Press.
Richerson, P.J. & Henrich, J. (2012): Tribal social instincts and the cultural evolution of institutions to solve collective action problems, *Cliodynamics: The Journal of Quantitative History and Cultural Evolution* 3: 38–80.
Rosenberg, A. (2017): Why social science is biological science, *Journal for General Philosophy of Science* 48: 341–369.
Rubin, H. (2023): Unlike agents: The role of correlation in economics and biology, in: du Crest, A., Valković, M., Ariew, A., Desmond, H., Huneman, P. & Reydon,

T.A.C. (Eds): *Evolutionary Thinking across Disciplines*, Cham: Springer, pp. 357–397.
Sawyer, R.K. (2005): *Social Emergence: Societies as Complex Systems*, Cambridge: Cambridge University Press.
Schaller, M., Gideon Conway III, L. & Peavy, K.M. (2010): Evolutionary processes, in: Dovidio, J., Hewstone, M., Glick, P. & Esses, V. (Eds): *The SAGE Handbook of Prejudice, Stereotyping and Discrimination*, London: SAGE Publications, pp. 81–96.
Schatzki, T.R. (2001): On sociocultural evolution by social selection, *Journal for the Theory of Social Behavior* 31: 341–364.
Scholz, M. & Reydon, T.A.C. (2008): The population ecology programme in organization studies: Problems caused by unwarranted theory transfer, *Philosophy of Management* 6: 39–51.
Stern, C. & Madison, G. (2022): Sex differences and occupational choice: Theorizing for policy informed by behavioral science, *Journal of Economic Behavior & Organization* 202: 694–702.
Sunsay, C. (2023): A historical evaluation of the disease avoidance theory of xenophobia, *PLoS One* 18: e0294816.
Tomasello, M., Melis, A.P., Tennie, C., Wyman, E. & Herrmann, E. (2012): Two key steps in the evolution of human cooperation: The interdependence hypothesis, *Current Anthropology* 53(6): 673–692.
Tooby, J. & Cosmides, L. (1989): Evolutionary psychology and the generation of culture, part i: Theoretical considerations, *Ethology and Sociobiology* 10: 29–49.
Tooby, J. & Cosmides, L. (1992): The psychological foundations of culture, in: Barkow, J.H., Cosmides, L. & Tooby, J. (Eds): *The Adapted Mind: Evolutionary Psychology and the Generation of Culture*, New York: Oxford University Press, pp. 19–136.
Turchin, P. (2018): *Historical Dynamics: Why States Rise and Fall*, Princeton: Princeton University Press.
Valković, M. (2023): Ontological and methodological limitations of certain cultural evolution approaches, *Philosophy of the Social Sciences* 54: 279–301.
Van Bouwel, J. & Weber, E. (2008a): A pragmatist defense of non-relativistic explanatory pluralism in history and social science, *History and Theory* 47: 168–182.
Van Bouwel, J. & Weber, E. (2008b): De-ontologizing the debate on social explanations: A pragmatic approach based on epistemic interests, *Human Studies* 31: 423–442.
Wartenberg, T.E. (1990): *The Forms of Power: From Domination to Transformation*, Philadelphia: Temple University Press.
Wilson, E.O. (1975): *Sociobiology: The New Synthesis*, Cambridge: Harvard University Press.
Zahle, J. (2016): Methodological holism in the social sciences, in: Zalta, E.N. (Ed): *The Stanford Encyclopedia of Philosophy* (Summer 2016 Edition), http://plato.stanford.edu/archives/sum2016/entries/holism-social/.
Zahle, J. & Collin, F. (Eds) (2014): *Rethinking the Individualism-Holism Debate: Essays in the Philosophy of Social Science*, Cham: Springer.
Zahle, J. & Kincaid, H. (2019): Why be a methodological individualist?, *Synthese* 196: 655–675.
Ziman, J. (1999): Marriage of design and selection in the evolution of cultural artefacts, *Interdisciplinary Science Reviews* 24: 139–154.

3 Cultural Evolution and Naturalized Social Ontology

Richard Lauer

3.1 Introduction

Metaphysics is the study of what the world is like, that is, what its fundamental structure is. Ontology is the part of metaphysics concerned with what there is in the most general sense. Given this, one can reasonably infer that *social* ontology is a field concerned with what there is in the social domain, and social metaphysics is concerned with, ultimately, what the social world is like. But how do we (or how should we) go about answering these questions? One approach, which I will champion in this chapter, is that we should pursue social ontological projects in a manner continuous with the empirical results of the social sciences. Specifically, I will argue that social ontological projects should rest on empirical results that are inferred using methods that are the most *discerning* of the methods available. I will detail this more later, but the notion of "discerningness" that I offer restricts the range of empirical studies from which we should draw lessons about the metaphysics or ontology of the social world.

This thesis has implications for the attitudes we should have toward the results of theorizing in cultural evolution. I will separate two (non-exhaustive) types of inquiry we find in studies of cultural evolution. First, there are what I will call *modeling-based studies*, where the goal is to provide an abstract model that describes how some social practice might have emerged. Second, there are *empirical studies* where the aim is to find a way to test hypotheses about the emergence of a social practice from operationalized concepts. Given my account of discerningness, if we want to know what the social world is like, we ought to rely on the results of empirical studies when we pursue metaphysical or ontological projects.

Consequently, my discussion contributes to this volume in two ways. First, it situates cultural evolutionary studies within a metaphilosophical debate about how to best pursue metaphysical (specifically ontological) projects (it is, consequently, "metametaphysical"). In doing so, this chapter illustrates the way in which cultural evolutionary studies might contribute

DOI: 10.4324/9781003643845-4

to answering metaphysical questions about the nature of social reality. In particular, as I discuss in Section 3.4, I argue that naturalistic approaches yield similar insights to those attempted by more traditional metaphysicians. Further, by providing suitably severe constraints (by way of the notion of "discerningness" I discuss later on), I offer an approach to social ontology that is in dialectical tension with naturalistic approaches to social ontology articulated in the literature.[1] Moreover, my approach emphasizes focusing on empirical work, where much mainstream literature in the philosophy of science focuses on abstract, usually mathematical, models (see Kincaid, 2024), where much attention is given to abstract modeling considerations rather than empirical considerations.[2] Finally, the results of my discussion point us to questions about which models in cultural evolution we should adopt a more realist stance toward, and which we should adopt a less realist stance toward.

My discussion will proceed as follows. Section 3.2 details some of the issues around "naturalizing" social ontology and lays out and motivates the main commitments of the brand of naturalized ontology that I will advance in this chapter. Section 3.3 then turns to distinguish two types of inquiry that we find in cultural evolutionary studies, which I dub the *modeling-based* and *empirical studies*. I give examples of each, drawing on the literature in cultural evolution. Section 3.3.1 goes over game theoretic models, like those formulated by O'Connor (2019), in which social practices (like categorizing individuals as belonging to particular genders) are the result of game theoretic situations. Section 3.3.2 goes over a paradigmatic instance of an empirical study – I discuss Holden's and Mace's (2003) attempts to operationalize and test hypotheses about the evolution of unilineal social structures. In Section 3.4, I visit some broader themes and implications of my approach. First, I visit the kind of scientific model that is best for drawing ontological inferences. Second, I argue that given the kinds of inferences we can draw from empirical models in cultural evolution, we can reject ontological individualism, and, furthermore, can do so in a manner that is informative of the broader dialectic – I visit Epstein's (2015) rejection of ontological individualism and compare my naturalistic argument to Epstein's.

3.2 Naturalized Social Ontology

The aim of any ontological project is to supply us with the means by which to infer that something or other exists. So, we should, upon arriving at a sound social ontological argument, have the basis by which to say that something is a part of the social world – that is, we should be able to say that some collection of individuals count as a group agent, or that, perhaps (or perhaps not), facts about society bottom out in facts about individual people. Let "naturalism" (as is a common use) be the position that metaphysics is (or should be) an entirely a posteriori discipline (Morganti, 2024). If it is an a posteriori discipline, it must locate our ontological commitments in the most reliable sources of knowledge that we have. Several philosophers endorse some version of naturalism so construed (Kincaid, 2024; Lauer & Khalifa, 2023; Little, 2021; Lohse, 2021; Ross, 2023). Naturalists do not accept that questions about the ontology of the social sciences should be settled by appeal to metaphysics. In the social domain, we might, in this case, appeal to the social sciences, just as we might appeal to the natural sciences when we endorse naturalism about metaphysics more broadly (Bryant, 2024; Ney, 2012).

What would it mean to "appeal to the social sciences"? To perform ontological inferences, naturalists rely on the best (available) scientific theories – it is in this sense that metaphysics is "a posteriori", since future empirical results may falsify our current metaphysical picture. The inferences drawn by naturalists constitute nothing over and above what is posited by first-order scientific theories (Ross, 2023). Given that the relevant ontological commitments come from the posits of first-order scientific theories (or in this case, social scientific theories), naturalists draw ontological inferences on the basis of the successful use of different tools used in the (social) sciences. For example, investigation of agent-based models reveals that there are structural assumptions in these models (Kincaid, 2024) that, if absent, would render these models unable to explain their target phenomena. If agent-based models enter into successful explanations, then we can (by naturalist standards) infer the existence of social structures. Moreover, some have argued that races are real because of the role they play in causal explanations in the social sciences (Kincaid, 2016, 2018; Mallon & Kelly, 2012; Root, 2000). The basis for such arguments is statistical models that reveal significant correlations between race and variables like educational attainment or socioeconomic status. In sum, the appeal to the social sciences is an appeal to the contents of well-supported models or theories in the social sciences.

Thus, science and ontology are tightly intertwined on the naturalist view. But it is worth noting just how tightly intertwined they are – let us do this by setting the contrast with some of the standard approaches in metaphysics that also seem to court naturalism. In metaphysics, a common assumption is *content naturalism* (Emery, 2023). Content naturalism is the view that the content of our metaphysical theories should be consistent with the content of our best scientific theories. It is worth highlighting what content naturalism allows – metaphysicians are free to posit stuff that goes beyond the first-order posits of our scientific theories. However, the addition of these commitments should not entail the falsehood of any of the claims of our best science. Naturalism does not let us go this far – we are locked into the posits of our best models.[3] Anything beyond what our best models (suitably interpreted) are committed to would not be justified by the evidence.

3.2.1 Discerningness

Now that I have given a general description of naturalism, I will sharpen this characterization and motivate a criterion we might use for naturalism, which I will call *discerningness* (Lauer & Khalifa, 2023). Recall, my aim will be to apply this to two kinds of inquiry in cultural evolution (which I detail in Section 3.3).

The core intuition behind my proposal is this: we should adopt the most probative standards we can if we want a body of scientific knowledge to serve as the basis for an ontology of the social world. To put things in Bryant's (2020) terms, we want constraints on ontological theorizing that are suitably *robust*. Constraints on ontological theorizing are robust to the extent that they admit theoretical content into an ontology, where more robust constraints exclude more theoretical content from an ontology. Consider, for example, that the demand for "continuity with science" might rule out supernatural entities from an ontology. Or, perhaps we adopt a materialistic stance as a constraint on metaphysical theorizing – then, abstract entities (like universals) might be prohibited from entering an ontology. *Discerningness* plays a similar role. If we should rest our social ontologies on the results produced through the use of our most discerning methods, we should rest our ontological commitments on social scientific claims that have the right kind of support, namely support from empirical studies. Unlike the examples mentioned, discerningness imposes a robust constraint on an ontology by restricting the range of methods we can use for supporting ontological commitments.

Discernment, as I am thinking about it here, is a relational property between methods, that is, one method is more discerning than another under certain conditions – for one method to be more discerning than another, we must

1. Specify methods *x* and *y* for answering question *q*.
2. Identify paradigmatic answers to *q* of which *y* counsels acceptance.
3. Show that some of the answers in Step 2 are competitors.
4. Show that *x*, unlike *y*, provides a reliable means of adjudicating between the competitors in Step 3.

Crucially, we need these methods to "talk" to each other, in the sense that they need to at least play a role in answering the same type of question. Consider, in anticipation of the discussion to come, that the common question might be about the origins of a particular set of social practices. From there, we should consider the different methods that might be used to answer these questions. I will later distinguish two such methods for the case of cultural evolution: there are instances where the aim of cultural evolutionary study is to provide an abstract, mathematical representation of a system that describes its possible evolution, and there are other instances where the aim is to mine the world for empirical evidence that would vindicate models of cultural evolution. Given Steps 2–4 above, if one of these is more discerning than the other, then for some answers to certain questions, *either* the empirical methods can adjudicate competitor answers generated using modeling methods, *or* things will go the other way round; modeling methods adjudicate competing answers to questions given by empirical methods. The central intuition underlying all of this is that the more discerning method provides the conceptual resources needed to adjudicate the different hypotheses, while less discerning methods cannot do this (it may be, e.g., that we have two potential accounts of a phenomenon with no means of ensuring that one is better than the other).

Step 4 presents a particular burden – in effect, showing that we can execute Step 4 of the process shows that we have at least some reason to believe that one method is a more reliable guide to truth than the other, since it can provide further reasons for accepting one answer over another. Notice how this relates to robustness: even if both of these methods are acceptable methods in scientific research areas, the reliability of one method over another should indicate to us which pool of scientific research should prop up our ontological commitments. So, while naturalism broadly construed says that our ontological commitments should rest ultimately on scientific results, discerningness is a criterion that further narrows the space of results that can be used in this way – that is, discerningness is sufficiently robust that it does not admit commitments from all of our best science.

But why think that we should adopt such a robust constraint? One of the main reasons is that we want to endorse claims that are likely to be true. Some scientific methods are likely to be better suited to this than others. Some, for example, might yield judgments that are *epistemically risky* in that they are judgments that are not empirically vulnerable (Chakravartty, 2017). For example, we might end up with ontological commitments to entities that we cannot possibly detect in the world – for which scientists are not able to set up an appropriate study (experimental or otherwise) to test, for example, abstract entities. In such cases, when we bet on the existence of these things by forming a commitment, we are taking on additional risk in forming false beliefs. When, by our best evidence, a claim about the world's contents is empirically vulnerable but has survived critical scientific scrutiny, we take less of a risk when we form ontological commitments. This is because, for naturalists, testing our claims against experience is a significant test of the truth of that claim. If it is the probative nature of scientific methodology that justifies the appeal to science for establishing ontological commitments, and it is that probative nature that promotes the reliability of a method, then discerningness would have us give preference to methods that yield commitments that are maximally empirically vulnerable, but ultimately less epistemically risky.

3.3 Modeling and Empirical Inquiry in Cultural Evolution

The aim of this section is to present two instances of research in cultural evolution with the aim of setting the stage for the application of discerningness to them – I argue, per the discussion of Section 3.2, that the ontological commitments we extract from some methods are more epistemically vulnerable, and so less risky than the ontological commitments we extract from other methods. In turn, we learn which cases it may be best to lean on if we want to construct ontological inferences about the contents of the social world using cases from cultural evolution.

I will use two cases of cultural evolutionary study that highlight some of the main research methodologies deployed in this area. I distinguish between two types of studies, and they cut across a wide range of the types of studies we find in cultural evolutionary studies, though these are not exhaustive of the types of studies that could fall under the mantle of "cultural evolution". The two studies are (1) modeling-based studies, where the aim is to construct an abstract, usually mathematical, model that describes the behavior of a system that undergoes cultural evolution, and (2) empirical studies, where the aim is not only to develop a model, but to find proxies or data that can stand in for variables in a model describing the evolution of social practices. I will discuss each approach in turn.

The basic structure of my argument proceeds along the following lines: first, there is a general question that both types of study are seeking to address, namely the question about what the best evolutionary explanation of a social practice is. In both cases, there are some candidate answers considered, but I focus largely on giving a sense of where contributors have disagreed. From there, I consider what these methods aim at – modeling methods aim to provide how-possibly explanations, while empirical methods aim to test theoretical claims against data. From there, I argue that the fourth step in the process of determining whether a method is more discerning than another is satisfied for empirical cases, but not for modeling cases.

3.3.1 Modeling the Evolutionary Origins of Social Categories

There are numerous models that have been used to give evolutionary explanations of society. For example, some cultural evolutionists have attempted to apply dynamical models in population ecology to patterns in human history (Turchin, 2003). In these studies, researchers use mathematical models to develop descriptions of the patterns of growth that might be expected – considerations like the size of a population, the availability of resources, and the chances of success at war are all considerations that might play a role in dictating the increase in groups of human populations, and as in the case of Turchin, what we might expect with respect to the growth and success of empires. There are also other models that one might try to develop, as, for example, in the case of O'Connor (2019) (discussed in Chellappoo's chapter in this volume), where researchers try to model how social categories like gender result from coordination games. To get a sense of how the modeling strategy works, let us consider a case in greater detail.

Consider two ways we might model the emergence of a gendered division of labor – a division of labor such that males typically take on some tasks, for example, hunting, while other females take on others, for example, caring for the young. On one approach, we might think that the gendered division of labor is the result of a certain kind of learning, namely by imitation. Males and females of a social group observe the behaviors of others around them and act accordingly. Of course, there are other mechanisms that might enforce these behaviors (males might be encouraged through social pressures to imitate males, and the same for females). The chief explanation, then, for the emergence of the gendered division of labor would be the result of a certain kind of learning process. This division of labor may then play into an account of the origin of gender categories (and perhaps similar stories may follow for others).

While imitation is no doubt a part of the process, some modelers have pushed back on the idea that the gendered division of labor is best modeled this way (O'Connor, 2019). Rather than assume that the gendered division of labor is the result of imitation and enforcement, it may be important to assume that some kind of social categorization already exists to facilitate the creation of such a division. O'Connor imagines a simple game where individuals are coordinating with each other, and two agents change their strategies in order to do better and do so using some marker that they come to recognize about each other. O'Connor writes,

> we should expect others to start imitating the behavior of these coordinators, by recognizing the same underlying marker and using it to condition behavior and so coordinate. Once more individuals start to recognize the marker, everyone using it to coordinate will do better than before, because they have more partners who know how to break symmetry.
> (2019: 97)

That is, we can imagine gender (the "marker") to facilitate the solution to a coordination game rather than emerging from the coordination game itself.

In a coordination game, players play simultaneously, and the successful strategies will be those that match the other player. For example, choosing which side of the street to drive down could be represented as a coordination game – players in this game benefit most when they both choose the same strategy (both choose to drive on the right side, and so avoid collision). So, we can imagine the players in O'Connor's imagined game using the marker as a way of bettering their circumstances in a similar way – O'Connor models the division of labor as a coordination game and argues that we can think of gender as a result of attempting to solve the problem of how to divide labor.

Let us return to my account of discerningness. We can say at least the following using the above discussion:

1. There are two competing (in the sense of being jointly logically inconsistent) answers to the question about the emergence of certain social categories, that is, gender. These are that gender evolved as a result of resolving coordination games, and that gender evolved *in order* to resolve the coordination game.
2. The modeling method does not supply ready tools for adjudicating which of these answers is best suited to answering these questions – in fact, those formulating these models acknowledge that these models are limited to giving "how-possibly" explanations about the phenomena.

In turn, the plausibility of these models is connected to observation and experimentation in fields relevant to cultural evolution, for example, developmental psychology, but the explanations produced cite only potential causes of the relevant phenomena (in this case, the construction of gender categories), and admittedly so. Otherwise, the acceptability of these models is based on their intuitive plausibility given patterns of behavior we might associate with the system in question.

Consequently, we receive at best *just-so* stories from model-based approaches. That is, these models are, seemingly, at best offering what might be termed "casual empiricism" – they claim explanatory victories without using the empirical data to support the models (Alexandrova & Northcott, 2013). Even where attempts have been made to defend the assumptions of cultural evolutionary models, it does seem there is reason to think that the assumptions have not been supported experimentally (Lewens, 2015). Of course, this is borne, to some degree, out of necessity. The systems studied by cultural evolutionists are complex and, moreover, reach into the distant past. It is not my goal to deny the value of these models, but instead to hedge our bets against their role in generating an ontology of the social world. I conclude only (by the end of this section) that modeling-based methods are not as discerning as empirical methods.

On the other hand, the empirical approaches I discuss next test models against (and, indeed, construct models from) empirical data about a target system. These, as we will see, fare better in terms of discerningness. I conclude that between the two types of method, we should consider the empirical types the more discerning.

3.3.2 *Empirical Studies of Unilineal Social Structures*

Let us see what else we can learn about social structure by appealing to the kind of naturalism I have articulated in this chapter. This section will discuss the practice of cultural phylogenetics, in which scientists attempt to perform phylogenetic inferences, that is, inferences about the evolutionary origins of cultural artifacts or practices. Often, phylogenetic inferences are represented through phylogenetic trees. In evolutionary theory, these trees aid in the explanation of the current distribution of traits in biological populations by inferring the evolutionary history of different populations, representing the nearest ancestors of some species, and, further, the common ancestors shared even by distantly related organisms. This practice goes at least as far back as Darwin's *Origin of Species*, which includes a diagram representing the common descent of current-day populations from past ones. If cultures evolve in the manner that biological populations evolve, then we should be able to perform phylogenetic inferences about cultural practices as well. Researchers have attempted to construct phylogenetic trees of different phenomena. Consider, for example, work by O'Brien et al. (2001), which represented purported evolutionary relationships between projectile points used by Paleoindians – the core idea is that the transmission of cultural information across generations would inform later attempts to form these points, and so it should be possible to construct a phylogenetic tree tracing lines of descent from past versions of projectile points to later.

The case I will focus on concerns the drawing of phylogenetic inferences about the origins of unilineal social structures (social structures that trace family lineage either matrilineally or patrilineally – i.e., either by tracing descent from females or males) using empirical models, informed by data about current groups of Bantu-speaking populations (Holden & Mace, 2003). Holden and Mace deployed language models showing the relationships between different variants of the Bantu language (per their report, the Bantu language comprises a group of 450 languages). They did this with the aim of testing whether the relationship between cattle ownership and patriliny is spurious or whether it reflects a real causal relationship. They argue for the latter – but note here, from the standpoint of discerningness, that we have two potential answers to a question about the origins of a social practice – and Holden and Mace propose phylogenetic trees as a way of answering this question. Just as phylogenetic trees might help to distinguish between cases of convergent evolution and those cases where species of organisms share a common ancestor, phylogenetic trees representing relationships of descent between different societies might yield insight into the relationship between cattle and methods for tracing lines of descent as either spuriously correlated or as coevolved characteristics of groups sharing common ancestry.

The tree representing the historical relationship between the different variants of the Bantu language also served as a representation of the spread of farming across central and southern Africa (Holden & Mace, 2003: 2426). The phylogenetic tree that results represents the prospective relationships of descent between the different Bantu-speaking groups in the region. Each of the present-day groups would have either a matrilineal or patrilineal system of inheritance – each group in the tree represents different language groups that were assumed to share a culture with each other. Their phylogenetic tree clustered these language groups according to the presence or absence of cattle and the rules of descent (patriliny or matriliny).

Holden's and Mace's central hypothesis was that the presence of cattle brings about patrilineal culture. In total, they collected data about 68 Bantu-speaking groups, documenting as well whether each of those 68 members had matrilineal or patrilineal methods of tracing ancestral lineage. Further, they recorded whether the groups possessed cattle. Using these data, they tested for the coevolution of cattle and matriliny and found a significant negative relationship between cattle and matriliny. Their interpretation of this is that livestock-based societies tend to emerge where there exists a marginal benefit to invest in sons rather than daughters. Rather, where there are greater benefits for investing in daughters, horticultural modes of production tend to be favored. With respect to the former, livestock are frequently used as marriage offerings, so investing in livestock to ensure better marriage prospects for sons would arise in those circumstances where the benefits of investing in sons are greater. By contrast, land inheritance rules that allow daughters to inherit land might provide the basis for matriliny since paternity uncertainty in sons (the uncertainty that the offspring of a son are, in fact, that son's offspring) would make inheritance less uncertain (Holden & Mace, 2003: 2431).

As a result, if a matrilineal society has cattle, either it moves away from a matrilineal structure to a patrilineal structure, or the cattle lose their value, and the matrilineal structure is maintained. Similar considerations apply to patrilineal societies with no cattle – these groups would either gain cattle and retain a patrilineal system of tracing lineage or eventually transition to a matrilineal structure. The probabilities calculated reflected the hypothesized phylogenetic relationships between the different language groups.

With all this said, let us turn to our naturalized ontology. Like before, let us take stock of the concept of discerningness:

1. There are two competing (in the sense of being jointly logically inconsistent) answers to the question about the emergence of certain social categories, that is, unilineal descent. These are that the relationship between patriliny and cattle is spurious, and that these are features that coevolved.
2. The empirical tools supply methods for adjudicating competing explanations – in fact, some of these studies were developed expressly to rule out certain possibilities, for example, ruling out spurious correlations.

If discerningness is an appropriate standard to use for determining whether the commitments that result from the use of a method should be admitted into an ontology, then there seems to be a prima facie reason to prefer the empirical style of inquiry – it supplies the better means of ruling out just-so stories about the evolution of social practices.

What sorts of ontological commitments are warranted by this? First, let us note the core elements of the reasoning and how they align with the general description I provided at the end of the previous section. The variables related to these discussions are (1) cattle and (2) matriliny or patriliny. The operationalization of these variables is straightforward: Holden and Mace look to ethnographic data about these groups The relevant object of interest would *not* be the cattle – we are not interested in adding cattle to the list of objects in our social ontology (these would fall under class of commonsensibles – we would no more refute the existence of cattle than we would the existence of any other garden-variety entity).[4] Rather, the element of interest is the *relationship* between cattle and social structure. That is, we want to figure out which processes exist to explain the existence of certain social structures. The processes in question are evolutionary in character – the basis for much of the reasoning Holden and Mace offer involves the possibility of representing cultural evolution in phylogenetic terms. So, not only might we add social structures to our ontology, but we may also add that some cultural processes have the structure of a phylogenetic tree.

This is, naturally, a controversial conclusion to draw. The very idea of conceiving of cultural evolution as having a tree-like structure has been suggested as problematic by some, though this is often pitched at a broad intuitive level. Consider Mesoudi's (2011) discussion of the application of phylogenetic methods to cultural evolution,

> Ideas, customs, practices, and technology, it is argued, often diffuse across cultural lineages due to horizontal transmission, making these lineages blend together.
>
> (Mesoudi, 2011: 99)

That is, while phylogenetic inferences in biological evolution are supported by the mechanisms associated with genetic inheritance, "inheritance" in cultural evolution frequently seems to admit of horizontal rather than vertical inheritance – cultural traits can be passed between cultural products, but in the case of biological evolution, species that result from distinct speciation events are not in a position to merge again.

My application of discerningness suggests that as long as phylogenetic structures result from the more discerning methods that we can apply to answering questions about cultural evolution, we should commit to them. General appeals to conceptual issues associated with thinking about cultural evolution along phylogenetic lines might raise some reason to avoid full-on ontological commitment. While space prevents a full discussion of this issue here, discerningness might only license ontological commitments in a provisional sense, a thin sense comparable to the way in which scientists might have endorsed the existence of caloric in the history of science. The caloric theory posited that caloric is a fluid, but empirical evidence suggested that the only way to appropriately account for heat was to make it infinitely rare. This is because if heat were a fluid, objects should have an increase in weight when heated, but experimental evidence suggested that this was not the case (Psillos, 1999: 117). Nevertheless, because it was the most worked out theory available at the time (the kinetic theory, in its youth, was not as well developed as the caloric theory), scientists maintained a committal, but hesitant attitude toward its existence. This suggests that a naturalized ontology might require only a thin sort of commitment – a commitment that takes us as far as the empirical evidence goes.[5]

3.4 Why Discerningness? Naturalism, Ontological Individualism, and Metaphysical Methodology

Given the discussion so far, we can return to the notion of discerningness and fit our examples into the procedure described in Section 3.2 for determining whether some method is more discerning than another. I sum up my points as follows:

1. There are modeling-based and empirical approaches to answering questions about some cultural evolutionary outcome.
2. Modeling-based approaches yield answers to questions about evolutionary processes, some of which are in competition with each other, but modeling approaches do not aim to adjudicate between different equally plausible models, but instead to provide "how possibly" explanations.
3. Empirical approaches (can) provide the means by which to adjudicate the dispute about which answer is best with regard to the question of

the evolutionary origins of social practices – and, in fact, theorists work to adjudicate different answers by showing that some pass empirical scrutiny, while others do not.

All of this suggests that we have a suitable basis for thinking that empirical approaches are more discerning than model-based approaches.

The result of this approach is that a certain kind of social scientific model yields ontological commitments that are empirically vulnerable and non-risky. These models contain certain variables X and some Y such that:

1. X is an independent variable.
2. The model includes some social outcome in the dependent variable Y.
3. Both X and Y must be operationalized.
4. X must make a statistically significant difference to the dependent variable of interest.
5. The model positing X and Y rests on methodologically sound assumptions.[6]

Such models will support ontological inferences, though the exact nature of the inference may differ depending on the research context. For example, some models are intended to show that the independent variable has a real causal effect that we should take seriously and attempt to interpret – race might be one such independent variable. On the other hand, social scientists might be interested in the nature of the relationship between the variables rather than the variables themselves. This may be the case where the variables in question represent objects whose existences are directly observable. Crucially, however, this depends on how the variables in the model are operationalized (whether the key explanatory variables are directly observed or, like in psychometrics, are latent variables).[7]

So, ultimately, discerningness pushes for acceptance of the commitments of statistical models developed in the social sciences. This has not been as much the focus of philosophers of science – much effort is invested in understanding the nature of the abstract models that social scientists use for understanding phenomena. But if this is right, then combined with Holden's and Mace's case study, we have some reason to think that privileging more discerning methods will yield useful insights for philosophers – in particular, I shall argue that this has some significant results for the debate about ontological individualism (the thesis that the social world is exhaustively determined by individual people). But appealing to the results of discerning methods will push away from ontological individualism. I shall turn to this now.

3.4.1 Discerningness and Ontological Individualism

One of the interesting consequences of my approach, and of my choice of case study in particular, is that the ontological commitments associated with Holden and Mace's study offer insights about the nature of the social world that have only been spelled out through the use of thought experiments in other areas of social ontology. Some, Epstein (2015), for example, have charged that non-individuals (non-agents) offer explanatorily valuable information, and ontological individualists (again: those who think that the social world is comprised entirely of individual people) cannot account for the role these things might play in explanation and prediction. First, I will detail Epstein's argument, since it is particularly illustrative, and then I will present why this story would be unsatisfying from the perspective of the naturalist.

Consider Epstein's Starbucks example. In this example, Starbucks goes insolvent because its machinery and equipment fail while its employees are asleep at home. Consequently, Starbucks goes insolvent. Starbucks' insolvency, if we are strict ontological individualists, should be a strict matter of the behaviors of individual people. However, it does not seem possible to explain Starbucks' insolvency using facts about the individual employees, since they were at home asleep when the machines broke down. This illustrates the falsehood of ontological individualism because it shows that the social world cannot be entirely determined by facts about individual people – we need to incorporate facts about material things that these individual people interact with as well.

Determination It is necessary that if G is the case, G grounds F.

That is, G determines F just in case the conditions for determination are satisfied. Ontological individualism is, purportedly, an assumption about the social world underlying all social science. The Starbucks case illustrates the following: individual people, that is, the Starbucks employees (G), do not ground Starbuck's insolvency (F). That is, it is possible that Starbucks goes insolvent even if no individual people are a part of the explanation. So, while a thoroughgoing ontological individualist about explanation might only posit the individual Starbucks employees to predict or explain Starbucks' insolvency, this leaves certain explanations unavailable.[8]

Returning to Holden and Mace, we can see that some of the same insights that Epstein makes come out of the discerningness-based approach to social ontology that I have provided in this chapter. Taking their model seriously, we can see a significant role for entities that are not individual agents in the explanation of the evolution of certain social practices. Given this, we have two approaches to yielding similar conclusions: one based on the insights drawn from cultural evolutionary studies, and another based on our intuitions about a hypothetical explanation.

3.4.2 *Discerningness and Metaphysical Methodology*

Arguments like Epstein's are undesirable by naturalist standards, since they seem to involve appeals to intuition, and the epistemic merit of philosophical intuitions is unclear (Ladyman & Ross, 2007). There are several features of these arguments that are worth noting, given the discussion so far:

1. These examples are highly stylized and describe no actual social scientific research context.
2. They appeal to philosophical resources to explicate their concepts. For example, grounding is a general metaphysical tool intended to help metaphysicians illuminate dependency relationships between levels of reality.
3. They offer some putative story for the success or failure of some scientific enterprise. For example, applying the grounding concept offers a diagnosis for social scientists' inability to predict facts like Starbucks' insolvency.[9]

Stylized examples like Epstein's are intended as a proof of concept for the thesis that pure metaphysical theorizing can add constructively to the social scientific toolkit. In doing this, Epstein's view yields a revisionary role for social ontology in relation to social science – while naturalists endorse the view that ontological commitments should come from the relevantly successful science, Epstein's argument illustrates that resources from outside of the sciences can help to correct the ontologies of those sciences.

Consider the following argument, given the above: suppose that some result from ontological investigation falls into conflict with some part of social scientific theory or the knowledge derived from that theory. If so, then given that ontological investigations may yield knowledge that would force revision of social scientific foundations, it would *not* be the case that incompatibility with our body of social scientific theory would count against that theory.[10] That is, it would be a mistake to reject the metaphysical theory that produced a result incompatible with the social sciences, since that incompatibility may be a consequence of *error* in the ontology of the social sciences. Such philosophers, therefore, seem to endorse a radically anti-naturalistic view about ontology. Social ontology does not need to focus on consistency with the results of the social sciences; it can make contributions to the social sciences – in fact, this view suggests that sound social ontological arguments may contradict claims of the social sciences and force revision of those claims.

Because of the role that this kind of view gives to metaphysics over science, naturalists would not find Epstein's reasoning convincing. The use of stylized examples and freedom from the burden of appealing to scientific research contexts would leave the appeal to criteria like discerningness ineffective, and would, furthermore, eliminate the need to use experience as the ultimate check against our metaphysical theories. But cases like Holden and Mace's illustrate the idea that the social world depends on a whole lot more than individual people *without* the need to appeal to such stylized examples. The evolutionary influences of cattle on rules of descent suggest that, indeed, in some cases, we need something other than individual people to explain social outcomes (here understood as either patriliny or matriliny). Land and cattle play a crucial role in Holden and Mace's explanation of the evolution of patriliny. But note further: contrary to the ambitious claims of Epstein, this insight demands no revision to the tools the social scientists used to make their claims – whatever the flaws of those tools, we have no reason to believe that claims like Epstein's reveal that there is anything deep about the insight that the social world is determined by more than individual people.[11]

So, note that the Naturalist would favor the argument drawing on Holden and Mace *at minimum* because it draws on the results of empirical models that deal with the specific disciplinary or research contexts of social sciences. But another point in its favor is that it seems to motivate some prima facie plausible arguments against ontological individualism as the correct ontology of the social world. This seems to follow straightforwardly from the argument of this chapter – since empirical methods are more discerning, and those commitments that result from applying those empirical methods contradict ontological individualism, the latter should be rejected. This is *not* necessarily true of those models that represent the modeling-based approaches – typically, those models assume only that there are agents acting and interacting to produce certain effects. If empirical methods are more discerning than modeling-based methods, then there is no call to infer that the social world must be comprised entirely of individual people – the social scientific record reveals otherwise.

Moreover, the appeal to the discerningness of empirical methods in cultural evolution offers another benefit. Appealing to discerningness diminishes the need to appeal to abstract, stylized examples when theorizing about the social world. While philosophers have paid varying degrees of attention to the details of work in the social sciences, much mainstream work seems to float free from the commitments of the social sciences. This has happened, I suspect, because social sciences are perceived to be less successful than the natural sciences, and perhaps that metaphysical and ontological inquiries are perceived as being in some sense *more fundamental* and *more general* than scientific inquiries, and so social ontologists take for granted that their inquiries concern the subject matter of the social sciences.[12]

3.5 Conclusion

I have argued that empirical studies in cultural evolution are more discerning than modeling-based methods. Discerningness is central to the way I want to approach thinking about drawing ontological inferences, that is, answering the traditional philosophical questions about what there is. Further, I have argued that there are some implications of my approach that are relevant to some broader themes in the social ontology literature – namely the view that metaphysical theorizing might make a substantial contribution to the social scientific toolkit. Given my argument, I have suggested that we can arrive at the same insights without having to concede a role to pure metaphysical theorizing in the sciences.

In total, my discussion shows that cultural evolution can aid in moving the dialectic around the methods of metaphysics forward. Once we take seriously the empirical methods used by cultural evolutionists for the explanation of cultural evolutionary phenomena, we can take on board ontological commitments that might have been omitted using modeling-based approaches as the basis for a social ontology. These particular ontological commitments allow us to meet the challenges of more traditional metaphysical methodologies – their justification comes from their role in a certain kind of empirical social science and so undercuts the role some have tried to articulate for purer forms of metaphysical theorizing.

Notes

1 Especially Kincaid (2024).
2 These trends are found in this very volume, see, for example, the chapters by Baraghith, Burman, and Chellappoo in this volume. Indeed, in the philosophy of science, modeling has received enormous attention. See, for example, Weisberg (2013) and Rice (2021). There are some treatments of statistical modeling in the social sciences, but these treatments are focused largely on questions of causation in the social sciences. See Russo (2009).
3 This does not preclude the possibility of disagreement about how to best interpret the results of our best models. So, not all philosophical issues are resolved by appealing to the contents of our best science.
4 Commonsensibles are the objects of everyday existence about which we tend to be at least naïve realists.
5 The idea that naturalized ontology might in some sense come apart from a strict scientific realism is discussed in Bryant (forthcoming), Bryant (2024), and Lauer (2022).
6 See Lauer and Khalifa (2023) for more about this account.
7 Consider Kincaid's (2016, 2018) naturalism. There he uses structural equation and latent variable modeling to show that we have reason to believe that there

exist races and ruling classes. Kincaid's approach is largely in keeping with my account of naturalism. The appeal to discerningness, consequently, yields an ontology that is similarly motivated to Kincaid. The relevant difference in this case is that discerningness explains why we might prefer validated constructs (they deploy the kinds of empirical methods that can better adjudicate between hypotheses).

8 Note: Epstein believes that this is a reflection of a deep-seated assumption in the social sciences, namely its anthropomorphism. I will not discuss this particular theme, but you can find some discussions and concerns about it in Lauer (2017).

9 There is another argument to be made, using similar terms, about the kinds of examples that have been used to motivate philosophical accounts of social structure (which rely on a particular metaphysics of social structure). Haslanger (2016) motivates an account of social structures as *graphs* using only stylized examples similar to Epstein's. However, a metaphysics of social structure can be motivated by appeal to social structure. For example, Lauer (2024) uses Blau (1977) to challenge accounts like Haslanger's. Discerningness might also be deployed here to motivate a different metaphysics of social structure from those derived using the types of stylized examples described above.

10 This argument demonstrates that an anti-naturalist social ontology falls into conflict with what Emery's (2023) content naturalism. According to content naturalism, we should not accept metaphysical views (of which ontological views are a subset) that fall into conflict with our best science. Suitably adapted to the social sciences, my reasoning suggests that anti-naturalists reject the view that social ontologies *must not* conflict with the theories and findings of the social sciences. This loosely in line with views like Trophardy (2024), whose anti-foundationalism about ontology allows ontology to be responsive to social science (esp. sociology).

11 Guala (2022) argues that social scientific practice does not vindicate Epstein's rejection of ontological individualism, since individualists may argue that it is the projectible properties of a social system that need to be understood individualistically. My discussion is compatible with this view – the cattle have no causal powers with respect to the evolution of patriliny or matriliny independent of the actions of the individual people who value them.

12 See Hawley (2018) for an example of this kind of skepticism.

References

Alexandrova, A. & Northcott, R. (2013): It's just a feeling: Why economic models do not explain, *Journal of Economic Methodology* 20: 262–267.

Blau, P. (1977): A macrosociological theory of social structure, *American Journal of Sociology* 83: 26–54.

Bryant, A. (2020): Keep the chickens cooped: The epistemic inadequacy of free-range metaphysics, *Synthese* 197: 1867–1887.

Bryant, A. (2024): Naturalized metaphysics without scientific realism, *Argumenta* 19: 13–33.

Bryant, A. (forthcoming): Social science and the naturalization of social metaphysics: Old biases and new advances, *Journal of Social Ontology*.

Chakravartty, A. (2017): *Scientific Ontology: Integrating Naturalist Metaphysics and Voluntarist Epistemology*, New York: Oxford University Press.

Emery, N. (2023): *Naturalism beyond the Limits of Science*, New York: Oxford University Press.
Epstein, B. (2015): *The Ant Trap: Rebuilding the Foundations of the Social Sciences*, New York: Oxford University Press.
Guala, F. (2022): Rescuing ontological individualism, *Philosophy of Science* 89: 471–485.
Haslanger, S. (2016): What is a (social) structural explanation?, *Philosophical Studies* 173: 113–130.
Hawley, K. (2018): Social science as a guide to social metaphysics?, *Journal for General Philosophy of Science* 49: 187–198.
Holden, C. & Mace, R. (2003): Spread of cattle led to the loss of matrilineal descent in Africa: A coevolutionary analysis, *Proceedings of the Royal Society of London* 270: 2425–2433.
Kincaid, H. (2016): Debating the reality of social classes, *Philosophy of the Social Sciences* 46: 189–209.
Kincaid, H. (2018): Debating the reality of race, caste, and ethnicity, *Philosophy of the Social Sciences* 48: 139–167.
Kincaid, H. (2024): A naturalist approach to social ontology, *Synthese* 203: 27.
Ladyman, J. & Ross, D. (2007): *Every Thing Must Go: Metaphysics Naturalized*, New York: Oxford University Press.
Lauer, R. (2017): Predictive success and non-individualist models in social science, *Philosophy of the Social Sciences* 47: 145–161.
Lauer, R. (2022): Motivating a pragmatic approach to naturalized social ontology, *Journal for General Philosophy of Science* 53: 403–419.
Lauer, R. (2024): Macrostructural explanation in the social sciences, *Synthese* 204: 95.
Lauer, R. & Khalifa, K. (2023): Should ordinary race talk be ontologically privileged? Moving social science into the philosophical mainstream, *Synthese* 202: 1–26.
Lewens, T. (2015): *Cultural Evolution*, New York: Oxford University Press.
Little, D. (2021): Social ontology de-dramatized, *Philosophy of the Social Sciences* 51: 13–23.
Lohse, S. (2021): Ontological investigations of a pragmatic kind? A Reply to Lauer, *Philosophy of the Social Sciences* 51: 3–12.
Mallon, R. & Kelly, D. (2012): Making race out of nothing: Psychologically constrained social roles, in: Kincaid, H. (Ed): *The Oxford Handbook of Philosophy of Social Science*, New York: Oxford University Press.
Mesoudi, A. (2011): *Cultural Evolution: How Darwinian Theory Can Explain Human Culture and Synthesize the Social Sciences*, Chicago: University of Chicago Press.
Morganti, M. (2024): *Metaphysics and the Sciences*, New York: Cambridge University Press.
Ney, A. (2012): Neo-positivist metaphysics, *Philosophical Studies* 160: 53–78.
O'Brien, M.J., Darwent, J. & Lyman, R.L. (2001): Cladistics is useful for reconstructing archaeological phylogenies: Palaeoindian points from the southeastern United States, *Journal of Archaeological Science* 28: 1115–1136.
O'Connor, C. (2019): *The Origins of Unfairness: Social Categories and Cultural Evolution*, New York: Oxford University Press.
Psillos, S. (1999): *Scientific Realism: How Science Tracks Truth*, New York: Routledge.
Rice, C. (2021): *Leveraging Distortions: Explanation, Idealization, and Universality in Science*, Cambridge: MIT Press.

Root, M. (2000): How we divide the world, *Philosophy of Science* 67: S628–S639.
Ross, D. (2023): Scientific metaphysics and social science, *Synthese* 202: 146.
Russo, F. (2009): *Causality and Causal Modeling in the Social Sciences*, Dordrecht: Springer.
Trophardy, Y. (2024): Ontological anti-foundationalism in sociology, *Philosophy of the Social Sciences* 54: 151–167.
Turchin, P. (2003): *Historical Dynamics: Why States Rise and Fall*, Princeton: Princeton University Press.
Weisberg, M. (2013): *Simulation and Similarity: Using Models to Understand the World*, New York: Oxford University Press.

4 Social Functions

Evolutionary and Non-Evolutionary

Frank Hindriks

4.1 Introduction

The social sciences harbor two conceptions of functions, evolutionary and non-evolutionary. The former plays a role in cultural evolutionary theory. In a manner analogous to evolutionary theory in biology, cultural evolutionary theory explains the reproduction and proliferation of so-called "cultural traits", such as the tendency to cooperate or the use of artefacts. Nowadays, non-evolutionary functions play a role primarily in equilibrium theories of institutions. Instead of their proliferation, they explain the persistence of institutions, such as money, marriage, and property. Thus, the evolutionary conception of a function provides the basis for what I call "reproduction explanations", and the non-evolutionary conception for "persistence explanations".

Function explanations face two problems, which concern the notions of a purpose and a value. Both of these notions play a central role in the commonsense notion of a function. For instance, hammers are used for the purpose of driving nails. As this is what they are good for, it explains why they are regarded as valuable. Now, it makes sense to attribute purposes to hammers because they are used by agents. However, cultural traits and institutions are not things that are used for a purpose. And, except perhaps for organizations, they are not agents. Furthermore, even if they have value, it is unclear how this could play a role in causal explanations. Due to considerations such as these, a different conception of a function is needed for social scientific explanations. What I call "the teleological problem" and "the axiological problem" consist of the challenges to conceive of functions without, respectively, the notion of a purpose and that of a value.

The main question of this chapter is whether and how functionalisms in the social sciences solve these problems. Rational choice functionalism is one of the few kinds of functionalism still defended within the social sciences.[1] It conceives of institutions as equilibria, which are the unintended consequences of individual behavior. Because of this, it does not suffer

DOI: 10.4324/9781003643845-5

from the teleological problem. However, it distinguishes institutions with a function from social structures without a function in terms of the amount of welfare they generate. I argue that, because of this, rational choice functionalism succumbs to the axiological problem.[2]

The second influential kind of social functionalism is cultural evolutionary theory. As I discuss in this chapter, this theory is heavily influenced by evolutionary theory in biology. Now, suppose that the latter has solved the problems just mentioned and that they have basically the same structure. Then, cultural evolutionary theory does not suffer from the teleological and axiological problems either. However, even if this is the case, it suffers from a third problem. It reduces culture to cultural traits, which are properties of individual agents whose behaviors are in equilibrium. This results in a rather austere conception of institutions (Valković, 2023).[3] According to what I call "the complexity problem", cultural evolutionary theory fails to capture the complexity of institutions. It does not do justice to the interdependences and interrelationships that exist between individual agents. Thus, the two most prominent kinds of functionalism in the social sciences face significant problems.

To try and solve them, I turn to a richer, hybrid account of institutions on which they consist not only of equilibria but also of social rules, which prescribe actions or give meaning to them.[4] I argue that the difference between social structures with and without a function can be explained in terms of the notion of a rule without recourse to values. So, a hybrid theory can solve the axiological problem that poses a threat to rational choice functionalism. Furthermore, if it incorporates the notion of a rule, cultural evolutionary theory can do more justice to the complexity of institutions. Thus, a hybrid theory of institutions can also solve the complexity problem.

After an introductory section (Section 4.2), I discuss the teleological and axiological problems: first, in relation to equilibrium theory (Section 4.3); and then in relation to cultural evolutionary theory (Section 4.4). The remaining sections discuss other problems that cultural evolutionary theory faces. Section 4.5 addresses the way cultural evolutionary theory conceptualizes information and motivation. Section 4.6 proposes a solution to the complexity problem.

4.2 Functions and Function Explanations

Social and technical artefacts have functions. The function of money is that of a means of exchange. And the function of a thermometer is to regulate temperature. These are the purposes for which these objects are used. Furthermore, they are valuable to those who use them because they can be used for these purposes. Thus, the commonsense notion of a function can be used to make sense of technological and social artefacts.[5] In other contexts, it is problematic to understand functions in terms of the purposes of values. The function of the heart is to circulate blood. However, the hearts do not adopt purposes. Furthermore, hearts beat involuntarily. Hence, there is no agent that uses it for a purpose or to realize a value. These considerations reveal that it is problematic to understand organs such as hearts in evaluative or teleological terms. It could even be taken to imply that hearts do not have functions at all.

John Searle (1995) argues that the notion of a function is inherently evaluative. And he rejects the idea that hearts as such have a function. The more popular alternative has been to distinguish between a teleological concept and a non-teleological concept of a function. In contrast to the former, the latter is naturalized and hence non-evaluative. However, proposals to explicate the non-teleological concept face two problems, "the teleological problem" and "the axiological problem".[6] The teleological problem is the challenge to analyze the concept without using the notion of a purpose. The axiological problem is the challenge to do so without the notion of a value.

To solve these two problems, Larry Wright (1973) proposed to explicate the non-teleological concept in terms of the role that functions play in causal explanations. He argued that, if an entity has a function, two things will be true: first, the function explains why the entity is there; and second, the fact that the entity is there explains why it performs the function. For instance, organisms have a heart because it pumps blood. Furthermore, the fact that an organism has a heart explains why its blood circulates. Together, the two claims account for the fact that hearts seem to have a purpose. After all, it continues to circulate blood, thereby keeping the organism alive. However, the way it functions can be explained without recourse to the notion of a purpose. Consider next the fact that, having a heart is, in some sense, valuable for an organism. This plays no role in the explanation either. In this way, Wright purges the ordinary notion of a function from teleological and axiological notions (see also Millikan, 1999).

Thus, a function explanation has two components, which are captured by what I call "the function schema":

(1) *X* does *C* and thereby brings about *E*.
(2) *E* causes *X* to do *C*.

The function schema is an analysis of the non-teleological or etiological concept of a function. This kind of function is commonly regarded as the manifestation of a disposition. If the function of *X* is to *E*, then *X* is disposed to bring about *E* by doing *C*. So, *C* is the cause of effect *E*. If *X* stands for a heart, *C* is the activity of pumping blood, and *E is* the fact that blood is circulated. And this fact (*E*) explains in turn why the heart continues to pump blood (*X* does *C*).[7]

The function schema applies to both evolutionary functions and non-evolutionary functions. Consider dowries. One question is why they have existed for so long. This question concerns the stability of the institution or its persistence. Another question is why and how this institution spread from one context to others. It pertains to its proliferation or, more technically, its reproductive fitness. Non-evolutionary function explanations account for the persistence of the entity; evolutionary function explanations account for its proliferation.

Furthermore, non-evolutionary function explanations, "persistence explanations" from this point onwards, involve a feedback loop: the effect causes the behavior that causes the next occurrence of the effect. In other words, the fact that *X* brings about *E* (by doing *C*) causes *X* to do *C*, which in turn generates the next occurrence of *E*. In this way, persistence explanations reduce the appearance of purposeful behavior to a self-reinforcing process. Persistence explanations also support a deflationary interpretation of the statement that having a function is good for an entity. This merely means that the effect reinforces the behavior that the entity displays, which contributes to its continued existence. In other words, the effect contributes to its persistence.

Evolutionary function explanations, or "reproduction explanations", account for the reproductive fitness of *X*. This means that they explain why there are other tokens of *X*. In this way, reproduction explanations account, for instance, for the proliferation of a species. Thus, reproduction explanations reduce the appearance of a purposeful development over time to reproductive fitness, which accounts for the spread of the behavior at issue. Finally, just as persistence explanations, reproduction explanations support a deflationary interpretation of the statement that an effect is beneficial for the entity that has the function: it contributes to its reproduction, or more precisely, it has been selected for.[8]

Thus far, I have distinguished two concepts of a function, a teleological and an etiological one. Furthermore, I have introduced two conceptions of etiological functions, an evolutionary and a non-evolutionary one. And I have discussed how they feature in reproduction and persistence explanations. Now, formulating some such explanation requires a theory. If it is to solve the teleological and axiological problems, such a theory has to fit the function schema, or something close enough. In light of this, I go on to ask, in Sections 4.3 and 4.4, whether equilibrium theory and cultural evolutionary theory (CET) do indeed fit the function schema.[9]

4.3 Non-Evolutionary Functions

Functionalism used to be very influential in the social sciences. But many functionalist schools of thought have perished (see endnote 1). The equilibrium theory of institutions is perhaps the most important exception. According to this theory, institutions are solutions to collective action problems. They facilitate coordination or enable cooperation. Coordination problems are problems of information. They involve two or more alternatives that are more or less equally attractive. So individual preferences align. Suppose individuals enact one alternative rather than the others. Then, all that is needed for them to continue to do so is that they expect each other to go on in the same way – and perhaps common knowledge of all this (Lewis, 1969; Binmore, 2008).

In contrast, cooperation problems are problems of motivation. They involve a partial conflict of interest, which has to be overcome if individuals are to cooperate instead of defect. Famously, this requires a transformation of the preferences individuals have (Ullmann-Margalit, 1977; Bicchieri, 2006). This can be due to a social norm and associated sanctions, which deter them from violating it. Another way in which cooperation can come about is if individuals regard the norm as legitimate and attribute weight to it because of this (Bicchieri, 2006; Hindriks, 2019). So, expectations are the key to solving coordination problems, and preferences are the key to solving cooperation problems.

Equilibria are self-reinforcing behavioral regularities. In the absence of significant perturbations, individuals continue to behave in the same manner. Given the preferences they have, they align their behavior with the existing regularity because they expect others to do so. Thus, there is a feedback loop from the regularity to the behavior of the participants, due to which the regularity continues to exist. In this way, equilibrium explanations qualify as persistence explanations, which support the attribution of functions to institutions. Their function is to generate coordinative or cooperative benefits (Schotter, 1981; Tuomela, 2002; Hindriks & Guala, 2021). This is the core claim of rational choice functionalism (RCF).

4.3.1 *The Axiological and Teleological Problems*

RCF solves the teleological problem in a convincing manner. Within the equilibrium approach, institutions are commonly conceived of as the unintended consequences of individual actions. So, they do not intend the institution to persist. Participants are not concerned with the institution as such. And even if they are, this is not why they comply. Instead, each maximizes her expected utility, treating the existence of the institution as a parameter in the equation. Thus, institutions persist due to the self-reinforcing dispositions of its participants. This reveals that the structure of equilibrium theory closely resembles that of the function schema.

In contrast, the axiological problem poses a substantial challenge to *RCF*. The reason for this is that, according to equilibrium theory, institutions are mutually beneficial. In light of this, it has been criticized for being too optimistic (Elster, 1989, 2015; Brennan et al., 2013).[10] As many institutions are harmful or unjust, this can seem to be a devastating criticism. However, the persistence of such institutions also requires explanation. Furthermore, *RCF* requires only that, if an institution has a function, individuals choose a strategy that results in a higher payoff than at least one of the alternatives they have. And this may well be the case even if the institution is harmful or unjust in some sense. Because of this, the optimism criticism is too superficial.

In this respect, it is instructive to consider a discussion of the issue within CET. Peter Richerson and Joseph Henrich (2012) argue that mechanisms such as sanctions and reputation can stabilize norms that do not benefit the participants of the institution, let alone the group as a whole. They qualify such norms as "maladaptive". According to this first argument, norms can be locally sub-optimal. Furthermore, they consider institutions of social dominance and point out that they can constitute equilibria that are "manifestly sub-optimal compared to other societies". Finally, the situation can be "so complex as to make it difficult or impossible to discover a path to a better equilibrium" (Richerson & Henrich, 2012: chapters 6–7).

Both arguments establish that things could be better. The first argument is local. It compares the actual situation with one without compliance mechanisms. In that other situation, individuals would not comply, and they would be better off because of this. The second argument is global. It compares the norm that is actually in force with norms that are supported elsewhere. In this case, the individuals would be better off if another norm were in force. Although these comparisons have evaluative significance, they have no explanatory relevance. Neither of them undermines the claim that, in the situation they face, participants benefit from complying with the norm. And that suffices to support the idea that it has a function.[11]

To unearth the real problem, we have to dig deeper. An equilibrium is Pareto-dominant exactly if, compared to another equilibrium, at least one agent is better off while none of them are worse off. *RCF* uses this notion to distinguish between equilibria with and without a function. Pareto-dominant equilibria have a function, whereas Pareto-inferior equilibria do not. Now, consider non-cooperative equilibria. They are self-reinforcing just as cooperative equilibria. This suggests that they have a function. However, proponents of RCF deny this.

The argument for this cannot have anything to do with the causal process that sustains the equilibrium because that is identical in both cases. The only difference is that non-cooperative equilibria do not solve a collective action problem. This means that, in contrast to cooperative equilibria, they do not increase the extent to which the participants satisfy their preferences. In other words, they are not Pareto-dominant. So, the only difference between cooperative and non-cooperative equilibria is the value they generate.

The upshot is that equilibrium theory solves the teleological problem. However, because of the role that Pareto-dominance plays in it, *RCF* fails to provide a deflationary interpretation of the sense in which the consequences of an institution are beneficial. It follows that, even though it fits something like the function schema, it fails to solve the axiological problem.

4.3.2 Rules and Equilibria

Yet non-evolutionary functionalism can be revived by enriching the theory of institutions. Elsewhere, Francesco Guala and I have argued that coordination institutions should be understood not only in terms of equilibria but also in terms of rules (Hindriks & Guala, 2015; Guala & Hindriks, 2015). Our key claim is that individuals converge on a common way of behaving by following signaling rules, which correlate their expectations. Social artefacts, such as traffic lights, military uniforms, and wedding rings, are signaling devices that facilitate coordination.

More recently, I have developed a similar argument for cooperation institutions.[12] Cooperation problems are frequently solved by means of normative rules that require cooperation (Bicchieri, 2006; Hindriks, 2022). These accounts imply that institutions consist not only of equilibria but also of rules. Those rules transform the games by making new equilibria available. In this way, they solve collective action problems. So, an institu tion is an interdependent regularity that is governed by a rule. That rule captures its symbolic or normative dimension. I call this "the Rules-and-Equilibria Theory" (Hindriks, 2025).[13]

So, how does this solve the axiological problem? It offers a new answer to the question of what distinguishes institutions with a function from those without. Institutions have a function because they consist in part of rules. Due to those rules, they have beneficial consequences. Those consequences are beneficial in the deflationary sense that they explain their persistence. They differ from non-cooperative social practices that do not involve rules. The participants of such practices do nothing more than repeatedly engage in the same activity. Because of this, they do not qualify as institutions (Hindriks, 2025).

According to the Rules-and-Equilibria Theory (RET), institutions are to be individuated in terms of their rules. They can remain the same even though the equilibria, to wit, the social practices they involve, change. Consider monogamous marriage. Initially, spouses are faithful to one another. But over time, many of them start cheating. Because it is still governed by the same monogamous rule, the institution as such remains the same. Yet, in one context, the rule is followed, whereas in the other it is frequently violated. In this respect, it changes.

What also speaks in favor of RET is that it supports a robust conception of malfunction. Well-functioning institutions are strong. Their participants have more than enough motivation to comply. In contrast, if an institution is weak, compliance is low (Hindriks, 2022). This means that the institution malfunctions. Thus, the notion of a malfunctioning institution can be explicated in terms of the attitudes that its participants have toward its rules. It performs its function if their attitudes support compliance. And it functions well if they have surplus motivation.

In sum, RET solves not only the teleological problem but also the axiological problem, in contrast to equilibrium theory. Because of this, the persistence explanations that it supports can plausibly be interpreted as non-evolutionary function explanations. The next question is whether CET offers adequate solutions to the teleological and axiological problems.

4.4 Evolutionary Functions

Cultural evolutionary theory (CET) is analogous to biological evolutionary theory (BET). In particular, both cultural and genetic evolution are processes of descent with modification. Some proponents of CET, such as Richard Dawkins (1976) and Alex Mesoudi (2011), claim that the analogy between the two kinds of theories is rather strong. If they are right about this, cultural traits can have evolutionary functions, just as biological traits. Instead of the reproductive fitness of biological organisms, they explain the proliferation of social entities. In the extreme, the only difference between them is the unit of selection: for instance, cultural traits instead of biological traits or, as I discuss later, memes instead of genes. If this is indeed the case, CET fits the function schema. This suggests that, insofar as BET solves them, CET can avail itself of the same solutions to the teleological and axiological problems.

However, in actuality, CET differs from BET in important ways. In the BET, the reproductive fitness of a species is explained in terms of a process of genetic inheritance, variation, and selection. Furthermore, genes, which provide the main focus of BET, encapsulate dispositions that account for the behaviors that increase the fitness of the species. However, whereas BET is fairly uniform, CET is extremely diverse.[14] Some theories that fall under this umbrella term invoke members, others cultural traits. Furthermore, some take cultural traits to be strategies of behavior, whereas others identify them with information. They form, respectively, "the behavioral approach" and "the informational approach", which also encompasses memes. I discuss them in turn.

4.4.1 *The Behavioral Approach*

Maynard Smith (Smith, 1982) adapted game theory for the purpose of evolutionary theorizing in biology. He proposed that selection does not work on individual traits in isolation, but in the context of a population. To analyze this, he proposed the notion of an evolutionary stable strategy, which captures the condition under which a behavioral strategy is stable within a population. As I discuss below, this approach readily allows for a functionalist interpretation. Inspired by Maynard Smith's theory, proponents of the behavioral approach started using evolutionary game theory to explain the proliferation of behaviors in populations of human beings.

To see how this works, consider Maynard Smith's explanation of territorial behavior among animals. In the hawk–dove game, which he proposed, two individuals compete over a resource. If both display hawkish behavior, they start by fighting and proceed to escalate. Those who behave like doves also start by fighting. However, if they encounter a hawk, they subsequently flee to safety. And if they meet another dove, they share the resource. Hence, a fight gets nasty only if two hawks meet. Depending on their initial distribution, one or the other strategy will increase until a stable state is reached, an evolutionary stable equilibrium. At that point, there is a mix of hawks and doves that is proportional to the value of the resource over the cost of losing a fight.

Within CET, such evolutionary games have been applied to cultural or institutional behavior, for instance, to property rights and military conflicts (Young, 2015, 1998). In these models, cultural traits are strategies in a game. Those strategies specify interdependent behaviors: the payoff of one agent depends on the strategies of the others. Furthermore, agents have fixed strategies, which are transmitted in an asexual manner. In order to spread within a population, a strategy has to form an equilibrium that is evolutionary stable. Such an equilibrium constitutes an institution. Thus, the function of that institution is to generate advantages for the individuals. These are the payoffs of the strategies, to wit, their cooperative benefits.

According to the behavioral approach, cultural traits are strategies or dispositions that realize them. On this interpretation, they are dispositions to perform interdependent actions, typically prosocial or even altruistic actions. For those dispositions to manifest themselves, they have to be in a context where enough others are disposed to display the relevant behavior as well. If those dispositions spread, the institution is reproduced. This approach is sufficiently analogous to BET so as to provide a viable conception of a social function. The reason for this is that, just as BET, it revolves around the notion of reproductive fitness. Because of this, it fits the function schema and solves the teleological and axiological problems. Let me elaborate.

The behavioral approach explains the reproductive fitness of a strategy in terms of the preferences that agents have in a particular context. It does not attribute a further reason for action to agents. In particular, it rejects the idea that agents purposefully reproduce the strategy. As it does not require recourse to the notion of a purpose, it solves the teleological problem. What about the notion of a value? Suppose a trait is good for an institution. On the behavioral approach, this means that the trait contributes to the reproduction of the institution. This implies that the trait has been selected for within the population. In this way, the statement that the trait is good for the institution is reduced to purely causal statements. It follows that the value of the institution, which is constituted by cultural traits, plays no role in the explanation. Hence, the behavioral approach solves the axiological problem as well.

4.4.2 *The Informational Approach*

Cultural traits are often explicated in terms of information (Mesoudi, 2011: 4; Lewens, 2015: 49). This informational approach postulates transmission processes, such as learning by trial-and-error, to explain how information spreads within or between populations. For this to happen successfully, information should be simple, observable, testable, and compatible with existing information (Lewens, 2015: 67). Furthermore, to be selected for, information must have a relative advantage over the status quo. As illustrations of this approach, I discuss Dawkins (1976) and Mesoudi (2011) because they aim to be maximally faithful to BET. Furthermore, Mesoudi also proposes a much-needed synthesis (see endnote 3).

Richard Dawkins (1976) introduced the notion of a meme, which is the cultural analogue of a gene. This view has become known as “memetics”. More recently, it has been developed further by Daniel Dennett (1996) and Susan Blackmore (2000). Memes are pieces of information. They encompass behaviors, ideas, skills, and artefacts. Think, for example, of catchphrases, fashions, melodies, religious beliefs, and technologies such as that of building arches. As memes are cultural traits, they do not reproduce sexually. However, just as genes, they are reproduced with high fidelity. They are, so to speak, contagious and move from one mind to another. This reproduction process allows for small mistakes, which generate variation. Furthermore, they are subject to a selection process that is similar to that of natural selection.

Yet the selection process that it postulates differs from natural selection in an important respect. The reproductive success of genes is explained in part by the proliferation of their hosts. So, biological traits with a function tend to contribute to the proliferation of the organisms they are part of. However, memes are known as “the viruses of the mind”. This is not only because they are contagious, but also because even successful memes do not always benefit their hosts. In fact, they can harm them. Memes induce behaviors, which explain why some memes fare better than others. Those behaviors determine their reproductive fitness. But this is only a matter of their own success, not that of their hosts. This is an important difference between memetics and BET.

In spite of this difference, memetics solves the two problems at issue here in basically the same way as BET. First, memes are not agents. As such, they do not form purposes. Second, even if they have value, this does not, as such, play a role in reproduction explanations. So, just as BET, memetics does without teleological and axiological notions.

Just as Dawkins, Mesoudi identifies cultural traits with pieces of information. They include "a multitude of beliefs, attitudes, preferences, knowledge, skills, customs, and norms", which individuals acquire "through social learning processes such as imitation, teaching, and language" (Mesoudi, 2011: 1). Traits can, for instance, spread due to "the intrinsic attractiveness of an idea, belief, practice [which] affects its probability of being acquired" (Mesoudi, 2011: 65). As another example, individuals with a preference for copying others with high social status or advantageous skills increase their chances of becoming prestigious themselves (Mesoudi, 2011: 73). In such ways, content- and model-biases affect their behavior.

Cultural traits include the use of artefacts, such as a bow and arrow. As they seem to have a teleological function, they require extra attention. Constructing and using a bow and arrow requires numerous conscious choices. Consider also processes of "guided variation", in which individuals "modify acquired cultural traits according to their own individual learning biases" (Mesoudi, 2011: 69). Such modifications are, presumably, purposeful. In response, the proponent of the informational approach can say two things. First, it is unproblematic to attribute purposes to individual agents. Second, the process of selection is still non-teleological. Individuals do not, or need not, aim at reproduction for it to occur.

Insofar as the axiological problem is concerned, Mesoudi (2011: 65) observes that it can be difficult to ascertain whether a trait is advantageous. But this may just be a theoretical problem. It seems still true that, if it is selected for, values play no role in the process. In light of this, I conclude that Mesoudi's version of CET solves both problems as well. Thus, the informational approach faces its own challenges. But there is no reason to believe that it cannot solve both the teleological problem and the axiological problem. However, it may still suffer from other problems. In the introduction, I mentioned the complexity problem, which I address in Section 4.6. But first, I investigate the notion of information that it employs and how information can affect behavior.

4.5 Information and Motivation

According to the informational approach, "culturally acquired information affects our behavior in quite fundamental ways" (Mesoudi, 2011: 1). But how? Information does not cause behavior on its own. It only does so if the agent has the requisite motivation. According to what I call "the motivation problem", the informational approach does not adequately explain how information affects behavior. Furthermore, it employs a conception of information that is too broad to be explanatory at all. It is not sufficiently unified to play a similar causal role in all of its applications. I call this "the information problem".

4.5.1 The Information Problem

Within the informational approach, culture is commonly taken to consist of "information capable of affecting individuals' behavior that they acquire from other members of their species through teaching, imitation, and other forms of social learning" (Richerson & Boyd, 2005: 5). It encompasses "knowledge, beliefs, attitudes, norms, preferences, and skills, all of which may be acquired from other individuals via social transmission and consequently shared across social groups" (Richerson & Boyd, 2005: 4). In a similar vein, Mesoudi takes "culturally acquired information" to include "beliefs, attitudes, preferences, knowledge, skills, customs, and norms" (2011: 1). These conceptions of information, or of their carriers, are rather broad.

The information problem is the challenge that they are too broad. On one conception, (correct) information is a representation of how things are. This holds for (true) beliefs. But not, for instance, for preferences and norms. On another conception, information concerns the way things might be. In other words, it consists of representations of actual or possible states of affairs. Yet preferences and norms do not represent possible states of affairs either. A preference ranking is an ordering of choices. And a norm specifies what individuals are to do. However, individuals can have expectations about each other's preferences or about the norms to which they subscribe.

So, information carriers can feature representations that concern how things actually are, including which attitudes individuals have, and how they could be or what is possible. I call them, respectively, "a-information" and "p-information". However, preferences and norms carry neither, at least not directly. Furthermore, insofar as the informational approach is meant to offer an alternative to the behavioral approach, information is distinct from behavior. Mesoudi makes this explicit when he claims that culture is "information rather than behaviour" (2011: 3).[15] Yet, just as many others, he includes skills, which can plausibly be seen as behavioral dispositions. In this respect, Mesoudi crosses the boundaries of his own conception of information.

Strikingly, customs and norms are also classified as cultural traits. However, such traits are attributes of individuals. And a custom is manifested in the behavior of many individuals. Furthermore, norms are in force only if they are suitably supported within a population. This suggests that they are the kind of things to be explained instead of those doing the explaining. Perhaps they can be explained in terms of cultural traits. But they are not cultural traits themselves. This supports the claim that the things counted as carriers of information are not sufficiently unified to play a similar causal role in all of its applications, which forms the core of the problem of information.

Tim Lewens proposes that the notion of cultural information is "best understood as an open-ended heuristic prompt which encourages an examination of the ways in which bodies of behaviours, skills, beliefs, preferences, and norms are reproduced from one generation to the next" (2015: 45). And if it is an open-ended heuristic prompt, any objection on the basis of what information really is becomes moot, it could be argued. However, a heuristic prompt for what? Presumably, for what can play a particular role in cultural evolutionary explanations. And for this purpose, the notion turns out to be too elusive.

Finally, as is well-known within CET, cultural information tends to blend during the transmission process (Lewens, 2015; Lewens & Buskell, 2023). However, it is vital to reproductive explanations that traits are reproduced with high fidelity. This observation threatens to undermine the analogy between CET and BET. And the solutions to the teleological and axiological problems discussed in Section 4.4 depend on this analogy. I do not mean to imply that CET cannot solve them. But it is unclear, at this point, whether and how it supports adequate reproduction explanations. So, the information problem poses a real threat to CET.

4.5.2 The Motivation Problem

Even given a coherent conception of it, the question remains how information influences behavior. According to the motivation problem, there is a gap in the explanations that the informational approach offers. It assumes that information can as such influence behavior. Yet it only does so if the individual has the requisite motivation. Suppose someone learns that it is raining outside. This information is completely irrelevant if she plans to stay inside all day. And if she plans to go outside, the way it affects her behavior depends on whether she wants to avoid getting wet or whether she wants to dance in the rain.

Beliefs carry a-information. If they are to explain behavior, they have to be complemented with desires. So, information supports behavior only against the background of preferences. And for it to support similar behavior among many members of a population, it has to be assumed that they have similar preferences. In Section 4.3, I discussed how coordination institutions persist due to preferences that are conditional on an expectation of a regularity in behavior. So, information about behavioral regularities can also support behavior. But in all these cases, preferences have to be presupposed.

P-information might concern desirable states of affairs. But how does such information contribute to the explanation of behavior? Suppose that, indirectly, preferences carry p-information. In that case, the answer is obvious. Preferences capture the motivation of the agent. Ironically, preferences need to be complemented with information if they are to explain behavior. This reflects back on the information problem. If preferences are carriers of information, they require further information as input to an explanation. It is far from obvious that this conceptualization supports a coherent action explanation.

4.5.3 Rules

The notion of a rule can help solve both of these problems, or so I propose. As it turns out, most proponents of CET rely on an equilibrium conception of institutions and conceive of them as regularities of interdependent behaviors (Lewens, 2015: 140). In Section 4.3, I distinguished between coordination and cooperation institutions. And I suggested that the collective action problems that they solve are, respectively, problems of information and motivation. Crucially, I proposed that signaling rules can solve the former, and normative rules the latter.

Coordination problems can be solved by means of signaling devices, such as police uniforms, traffic lights, and wedding rings. For individuals to respond to them, they must have the relevant signaling rules in mind. Such rules apply to interdependent situations in which individual preferences align. They tell them what to do when they encounter the devices. If they are generally complied with, everybody benefits. It follows that, if individuals expect others to comply, they will follow the rules themselves. In contrast, normative rules solve cooperation problems insofar as they transform individual preferences, for instance, due to associated sanctions. This serves to resolve the conflict of interest that exists in the situations to which they apply.

Both signaling rules and normative rules are interdependent rules for action. Such rules connect situations to behaviors. As such, the notion of a rule provides for a clear conception of the kind of information that is at issue. Furthermore, both kinds of rules feature in expectations about what other individuals believe and in individual beliefs. Those expectations and beliefs can be seen as carriers of information. Finally, if individuals comply with them, they behave accordingly. So, in combination with individual preferences, they explain behavior. In these ways, the notion of a rule can be used to solve the information problem as well as the motivation problem, at least to a large extent.

CET could treat interdependent rules, or the attitudes in which they feature, as traits that are reproduced in a population. They spread within and between populations where the situations to which those rules apply exist. In this way, they can feature in reproduction explanations. Signaling rules come with signaling devices, which are artefacts that, in some sense, carry information. They are reproduced as well. Cooperation is also supported by reputation, or undermined for that matter. This provides for another clear case in which information plays a role. Sanctions and perceived legitimacy support cooperation insofar as they lead to payoff transformations.

Interdependent rules imply p-information. If they are generally complied with, they result in a mutually beneficial behavioral regularity. The p-information is a representation of such a regularity. More immediately, interdependent rules encode what behavior is fitting in which situation. All in all, I conclude that the informational approach can at best provide a partial account of the reproductive fitness of cultural traits. Behavioral dispositions must be part of the story as well.[16] Thus, RET can help solve the information and motivation problems from which the informational approach suffers. In both respects, the notion of a rule provides for a plausible conception of cultural traits.

4.6 The Complexity Problem

According to the complexity problem, CET is not rich enough to adequately explain cultural phenomena. Culture does not just concern individuals, but also the interdependences and interrelations between them. Models that conceptualize it in terms of individuals with common traits cannot capture this complexity. They are too reductive (see endnote 3). Along these lines, Martina Valković (2023) has criticized Mesoudi's version of CET. She argues that he fails to solve the problem: Mesoudi does not "sufficiently appreciate the complexity of human sociality and culture" (Valković, 2023: 19). Here, I defend a similar conclusion. But I do so by more systematically exploring social ontology, the methodology of CET, and the relationships between them.[17]

To this end, I distinguish between a weak and a strong version of the complexity problem. According to the weak version, the complexity of social reality cannot be captured in terms of individuals and their cultural traits. This objection leaves open the possibility that a richer but still individualist theory can do so. In contrast, the strong version takes the complexity of social reality to support a non-reductive or holistic ontology. This implies that a reductive theory and methodology are not able to capture the complexity of social reality.

4.6.1 *A Non-Reductive Ontology*

CET relies on austere conceptions of institutions. As mentioned in Section 4.5, most of its proponents rely on an equilibrium conception of institutions. They conceive of them as regularities of interdependent behaviors. RET provides for a richer conception. This is in part because of the roles that interdependent rules play in it. As discussed in Section 4.3, rules can capture the symbolic and normative features of social structures. Furthermore, they can also be used to explain the significance that social roles have for institutions, as well as that of material resources (Haslanger, 2018; Hindriks, 2025). Equilibrium theories cannot do justice to this, at least not on their own. In this way, rules, roles, and resources give more substance to the claim that social reality is complex.

What is more, RET supports a non-reductive or holist social ontology (Hindriks, 2013, 2025). It does so in part by acknowledging the role that context plays in the constitution of institutions and institutional objects. This reveals the significance of interrelationships between institutions. There are at least two generic reasons for defending ontological holism. First, social entities can plausibly be taken to have causal powers of their own. For instance, purchasing power does not seem to inhere in pieces of paper or metal, but in the social objects they constitute. Individuals can use them as money only because of the wider context in which they find themselves, which includes a central bank and a state. Second, social entities can persist despite changes at the individual level. For instance, different individuals can, consecutively, occupy the same social role. And, usually, the role matters more than the individuals who occupy it (Hindriks, 2025).

As a social theory, RET reveals that CET suffers from the weak version of the complexity problem. When combined with a non-reductive social ontology, it backs the strong version of the problem.

4.6.2 *Ontology and Methodology*

Mesoudi is well aware of the weak version of the complexity problem: "many social scientists often argue that cultural change is too complex to be reduced to the simple mathematical models described earlier, such as neutral drift or population dynamical models" (2011: 131). This would mean that "culture is too complex to be analyzed using simple evolutionary models" (Mesoudi, 2011: 131). A related complaint is that, because of their complexity, "specific cultural phenomena (e.g., languages, empires) cannot be meaningfully compared with one another" (Mesoudi, 2011: 131). Yet this is exactly what proponents of CET try to do. In response, Mesoudi maintains that "the reduction involved is methodological" (2011: 131). So, he proposes what I call "a methodological solution".

On a strong reading, the claim is that the reduction is methodological rather than ontological. And if the ontology is not reductive, it is holist. This would mean that Mesoudi combines ontological holism with methodological individualism. But this makes little sense. As just mentioned, ontological holism supports social persistence and social causation. Presumably, this implies that social facts or events should also play a role in the explanation of social phenomena. Our understanding of them would be woefully inadequate without social-level explanations. In particular, the explanation of causal powers at the social level requires making reference to social phenomena. Because of this, ontological holism supports methodological holism.

On a weak reading, the reductive methodology is based on a reductive ontology. And the question is merely whether and how it adequately captures the complexity of social reality. In response, Mesoudi seems to use two strategies. The first is to insist that simple models can actually do so. BET has succeeded in capturing biological complexity in terms of simple models; CET can do so as well (Mesoudi, 2011: 131). Furthermore, there is not really an alternative, because only simple models support "quantitative testable predictions" (Mesoudi, 2011: 131). The alternative is to give up on fully understanding social reality, but this is "a rather pessimistic and intellectually sterile sentiment" (Mesoudi, 2011: 132). This is a striking way of putting the point, as the charge against him is that his models are too austere.

The second strategy he uses is that the charge reveals a "misunderstanding of the rationale behind using simple models" (2011: 132). He adds that "it is not being argued that human culture really can be described entirely in terms of just a handful of simple variables and processes" (2011: 132). This immediately precedes his claim that "the reduction involved is methodological" (Mesoudi, 2011: 131). But how can these two claims be squared with one another? Simple models may have virtues of their own. But if they fail to fully represent social reality, they cannot tell the whole story. So, without further explanation, these claims do not solve anything. As there are alternatives, I can only conclude that proponents of CET could try harder to enrich their theory and incorporate interdependences and interrelationships.

All things considered, I think that Mesoudi is best interpreted as being committed to both ontological and methodological individualism. In this respect, I agree with Valković (2023) who argues that his reductive methodology is motivated by his individualist ontology. Yet the views are independent. To see this, I first define the terms at issue (Watkins, 1959; Little, 1991; Elster, 2015):

Ontological individualism (OI): All social entities consist of individual agents and actions only.
Methodological individualism (MI): All social phenomena can ultimately be explained in terms of individual agents and actions.

Ontological holism (OH) is the negation of OI. And methodological holism (MH) is the negation of MI. Earlier in this section, I argued that OH implies MH. I go on to argue that OI does not imply MI.

According to MH, some social phenomena can be satisfactorily explained only in terms of other social phenomena. This leaves open the possibility that there are exceptions. So, it could be that some social phenomena can be adequately explained in terms of individual agents and actions. I call the view that there are indeed exceptions "methodological pluralism" (MP). A proponent of MP will explain some social phenomena in terms of individual agents and actions, and others in terms of social facts or events. Alternatively, she will explain one aspect of a social phenomenon in individualist terms, and others in more complex terms. It is a perfectly good option for proponents of CET who embrace OI to combine it with MP.

Valković claims that Mesoudi's "methodological assumptions . . . follow directly from his ontological ones" (2023: 279). But OI does not entail MI. To be sure, proponents of OI reject social causation. However, in spite of this, explanations that refer to social entities can still provide distinct causally relevant information. Consider the relationship between unemployment and crime. And suppose that, if there is an increase in unemployment, this leads to an increase in crime. Even if the causal relations occur at the individual level, a lot of information about individuals will be distracting. Think, for instance, of the various reasons why they become unemployed and why they engage in criminal behavior. Furthermore, it may be that the macro-relation is robust to variation at the individual level. So, it would remain the same even if the individual stories were different (Jackson & Pettit, 1992). It follows that OI is consistent with MP.[18]

Most of the time, Mesoudi comes across as a staunch individualist. He identifies culture with information. And he claims that "artifacts, speech sounds, and stated beliefs are the outward behavioral expressions of information stored in the brain and as such are the cultural equivalents of phenotypic traits such as height or skin color" (Mesoudi, 2011: 42). Furthermore, he explains cultural change in terms of content, model, and frequency biases. Finally, Mesoudi equates social structures with the distribution of cultural traits within populations. On the face of it, this commits him to a rather strong version of both OI and MI.

However, at some point, Mesoudi claims that "micro-level disciplines such as psychology have failed to acknowledge the extent to which macrolevel cultural processes shape individual behavior" (2011: 52). He adds that "macrolevel phenomena may be both caused by, and in turn influence, microlevel individual behavior" (Mesoudi, 2011: 226n81).[19] For instance, the frequency in a population can influence individual-level processes, as when someone exhibits a conformist bias. As it is unclear what he means by "influence", it would go too far to conclude that he is committed to social causation. So, it does not entail OH. At the same time, it does not seem consistent with MI. So, it may be that he should embrace MP even on his own terms.[20]

4.6.3 *Rules and Roles*

In Section 4.5, I argued that RET can help solve the information and motivation problems. Here, I propose that it can contribute to a solution of the complexity problem. For instance, social roles and institutional statuses can be explicated in terms of rules. A social role can plausibly be taken to consist of one or more rules that are generally acknowledged. It can remain the same even when it becomes occupied by a different individual. Furthermore, it gives the individual who occupies it powers he would not have otherwise. Yet he owes them in part to those who at least go along with the idea that he has them. So, the relationships that exist between the individuals are vital for understanding how social roles work (Hindriks, 2025).

Can the kind of models that Mesoudi works with capture this complexity? Valković argues that they cannot. She observes that models of the prestige bias assume that someone's prestige is inferred from her status, for instance, that of an art director. Furthermore, she argues that "it would seem that the socially assigned status, or rather a role, is what makes the difference in the experiment above, and it is unclear how Mesoudi's theory could accommodate this fact" (Valković, 2023: 296). In response, he might claim that it is built into the assumptions of the model. It would surely help if the narrative about the model were more explicit about this. Yet a model about the effect of statuses or social roles need not include an account of what they are.[21]

By way of analogy, consider the prisoner's dilemma in which two prisoners face a district attorney. They have a role that is defined by their context. Yet this is implicit in what are commonly referred to as "the rules of the game", which concern the situation that the agents are in and the features they have. Similarly, the models within CET postulate types of agents. Models of the prestige bias assume that some of them have statuses. Yet it need not be problematic if these assumptions are left implicit. This could be what Mesoudi has in mind when he claims that some misunderstand the kind of models he uses. Simple models can indeed be more powerful than one might think.

However, for this response to work, the account of social roles on which the model relies has to be adequate. In this respect, it is again useful to consider the prisoner's dilemma. It presupposes an account of what prisoners and district attorneys are. Suppose that these roles exist due to the relations they bear to other social phenomena as well as to the context in which they feature. On this supposition, the game theoretic model should be interpreted in holistic terms (Satz & Ferejohn, 1994). One might object to this because it revolves around individuals, their preferences, and expectations. But this objection fails to appreciate the role played by the rules of the game. The point I am making here is that, once that is fully appreciated, the game theoretic model is best interpreted non-reductively.[22]

Thus, I propose a dual-pronged solution. In some respects, the models themselves have to be changed and enriched; in other respects, their interpretation. To these ends, CET could use some of the insights RET provides, including the roles that the notion of a rule plays in it. Furthermore, CET should be combined with MP instead of MI and with OH instead of OI.[23]

4.7 Conclusion

The social sciences harbor two conceptions of social functions, an evolutionary and a non-evolutionary one. The evolutionary conception of a function supports reproduction explanations, which account for the proliferation of social entities. The non-evolutionary conception can be used to formulate persistence explanations, which make sense of their stability or resilience. I have argued, however, that the most influential contemporary theories that provide such explanations face significant problems.

Rational choice functionalism solves the teleological problem, but not the axiological problem. It distinguishes between institutions with and without a function in terms of their value. CET can, in principle, solve the teleological and axiological problems. But it tends to be rather reductive. It thereby risks falling foul of the complexity problem, in which case it fails to explain important aspects of social reality. This means that, as it stands, many if not most functionalist explanations in the social sciences are inadequate.

I have argued that these problems can be solved by means of the notion of a rule. First, according to the RET, institutions involve interdependent rules, which can, in principle, solve collective action problems. In this respect, they differ from ordinary social practices. Against this background,

I have proposed that institutions usually have functions, whereas social practices do not. This provides for a principled way of attributing functions to institutions that revolves around rules and does not invoke values. In this way, RET solves the axiological problem.

Second, many proponents of CET employ an austere conception of institutions as equilibria. Furthermore, they equate culture with information, which in turn encompasses anything from beliefs to preferences, and from customs to norms. This conception of culture is rather eclectic. And it is not always clear how it is meant to explain behavior, as it has little to say about motivation. Again, the notion of a rule can at least be part of the solution. Rules encode information. Furthermore, they connect situations to behaviors. In addition to this, they can capture the complexity of social reality insofar as they account for social roles and material resources, along with the persistence and causality of social structures.

All in all, the problems from which functional explanations suffer can be solved. This means that there are indeed two viable conceptions of function in the social sciences that can explain important features of social reality. The way in which I have defended these claims reveals how social ontology and CET can enrich each other. On the one hand, social ontology stands to benefit from CET, as it sheds light on the reproduction of social structures. On the other hand, the notion of a rule, as it features in social ontology, has great significance for CET.

Acknowledgments

I gratefully thank Richard Lauer, Simon Lohse, Thomas A.C. Reydon, and Martina Valković for their insightful comments.

Notes

1 Ullmann-Margalit (Ullmann-Margalit, 1977) was probably the first to explicitly defend rational choice functionalism. Earlier functionalist schools of thought in the social sciences include organic functionalism, neo-functionalism, structuralism, and Marxism. See Little (1991) and Bigelow (1998) for discussions.

2 For somewhat similar objections, see Elster (1989) and Brennan et al. (2013).

3 Lohse and Valković discuss two versions of this problem in Section 3.1 of Chapter 2. The first concerns culture and meaning, the second the reality of social structures. I focus here on this second version.

4 Such hybrid accounts are offered, among others, by Aoki (2001), Greif & Kingston (2011), and Guala & Hindriks (2015).

5 I do not mean to imply that the common-sense concept also provides for the best analysis of social and technical artefact functions. For a philosophical analysis of social artefacts and their functions, see Hindriks (2020). See Vermaas & Houkes (2003) for one of technical artefact functions.

6 In Hindriks (n.d.), I use the term "the value problem" instead of "the axiological problem".

7 A third influential concept of a function is that of a Cummins or causal role function (Cummins, 1975; Walsh & Ariew, 2013). Wouters (2003) distinguishes four kinds of function and argues that each of them play a central role in biology.

8 The difference between persistence and reproduction explanations is sometimes formulated in terms of the claim that the former concern tokens and the latter types (Jackson & Pettit, 1992; Pettit, 2007; Schulz, 2021).

9 Although I focus here on etiological functions, I should stress the significance of the teleological concept. For one thing, it is important for the analysis of social artefacts (Searle, 1995; Hindriks, 2020). Furthermore, etiological and teleological functions can be combined in a framework that bridges the gap between social science and social ontology on the one hand and policy and politics on the other (Hindriks & Guala, 2021). In Chapter 7, Åsa Burman and Nemi Pelgrom discuss teleological conceptions of normativity.

10 Although he rejects non-evolutionary function explanations, Elster (1989, 2015) accepts evolutionary function explanations in the social sciences.

11 So, Richerson & Henrich (2012) are too quick to qualify such institutions as maladaptive. In the technical sense of the term, they are not. In fact, it may be that, from a purely etiological point of view, the institution functions well.

12 I am concerned here with institutions that are not organizations. Elsewhere I extend this account to organizations as systems that promote centrally coordinated cooperation (Hindriks, 2025).

13 Guala and I call the part that concerns coordination institutions "the Rule-in-Equilibrium Theory" (Hindriks, 2019, 2022).

14 See Reydon (2023) for a rich and insightful discussion of BET, which has of course its own internal differences. For one of many reviews that reveal how diverse CET is, see Nichols et al. (2024). This claim about diversity is also supported by the fact that knowledge synthesis has been identified as one of the grand challenges of CET (Brewer et al., 2017).

15 Mesoudi (2011: 3) excludes behavior out of concerns for circularity. The worry is that behavior cannot explain behavior. Yet genes explain their own proliferation. See Lewens (2015: 49–50) for a more elaborate response to this worry.

16 Guala (2019) argues that preferences are best conceived of as behavioral dispositions.

17 In contrast to the information and motivation problems, the complexity problem applies not only to the informational approach but also to the behavioral approach.

18 Some individualists deny social causation, but acknowledge that interdependences and interrelationships play a more important role than the individualist is willing to grant. Carter (1990) calls this view "interrelationalism". This term is useful for those who are hesitant to embrace MP because it allows for holistic explanations.

19 Mesoudi makes a perceptive observation when he claims that "the complexity of this circularity may be one of the reasons why the micro-macro divide has persisted for so long in the social sciences" (Mesoudi, 2011: 226n81). Yet, it is difficult to see how an individualist account can do justice to the kind of feedback processes at issue.

20 In spite of the language Mesoudi uses, Valković (2023: 283–284) claims that the kind of models at issue can readily be interpreted in individualist terms.

21 For more on the prestige bias, see Chapter 6 (Baraghith and Jordan).

22 Van Hees (1997) recognizes that the rules of the game present an obstacle for a reductive or individualist interpretation of its methodology. He argues,

however, that the rules of the game can themselves be explained reductively. As he recognizes, this requires a reductive explanation of the rules of the next game, and so on. But if relational and contextual factors are relevant in each game, the holist will conclude that a non-reductive interpretation is inescapable.

23 Richerson & Henrich (2012) regard institutions as population-level phenomena that emerge from individual decisions, interactions, and learning. Although this sounds less individualist than Mesoudi's proposal, how exactly their view is to be interpreted depends, for instance, on what they mean by "emerge". Even if they would embrace MP, I see no reason to believe that they would accept OH. However, if the argument in the main text is successful, they should do so.

References

Aoki, M. (2001): *Toward a Comparative Institutional Analysis*, Cambridge, MA: MIT Press.

Bicchieri, C. (2006): *The Grammar of Society*, Cambridge: Cambridge University Press.

Bigelow, J. (1998): Functionalism in social science, in: *Encyclopedia of Philosophy*, London: Routledge, pp. 813–819.

Binmore, K. (2008): Do conventions need to be common knowledge?, *Topoi* 27: 17–27.

Blackmore, S.J. (2000): *The Meme Machine*, Oxford: Oxford University Press.

Brennan, G., Eriksson, L., Goodin, R.E. & Southwood, N. (2013): *Explaining Norms*, Oxford: Oxford University Press.

Brewer, J., Gelfand, M., Jackson, J.C., MacDonald, I.F., Peregrine, P.N., Richerson, P.J., Turchin, P., Whitehouse, H. & Wilson, D.S. (2017): Grand challenges for the study of cultural evolution, *Nature Ecology & Evolution* 1: 0070.

Carter, A. (1990): On individualism, collectivism and interrelationism, *The Heythrop Journal* 31: 23–38.

Cummins, R. (1975): Functional analysis, *Journal of Philosophy* 72: 741–765.

Dawkins, R. (1976): *The Selfish Gene*, Oxford: Oxford University Press.

Dennett, D.C. (1996): *Darwin's Dangerous Idea: Evolution and the Meanings of Life*, London: Pinguin.

Elster, J. (1989): *Nuts and Bolts for the Social Sciences*, Cambridge: Cambridge University Press.

Elster, J. (2015): *Explaining Social Behavior: More Nuts and Bolts for the Social Sciences*, Cambridge: Cambridge University Press.

Greif, A. & Kingston, C. (2011): Institutions: Rules or equilibria?, in: Schofield, N. & Caballero, G. (Eds): *Political Economy of Institutions, Democracy and Voting*, Dordrecht: Springer, pp. 13–43.

Guala, F. (2019): Preferences: Neither behavioural nor mental, *Economics and Philosophy* 35: 383–401.

Guala, F. & Hindriks, F. (2015): A unified social ontology, *The Philosophical Quarterly* 65: 177–201.

Haslanger, S. (2018): What is a social practice?, *Journal of Philosophy* 82: 231–247.

Hees, M. van (1997): Explaining institutions: A defence of reductionism, *European Journal of Political Research* 32: 51–69.

Hindriks, F. (2013): The location problem in social ontology, *Synthese* 190: 413–437.

Hindriks, F. (2019): Norms that make a difference: Social practices and institutions, *Analyse & Kritik* 41: 125–146.

Hindriks, F. (2020): How social objects (fail to) function, *Journal of Social Philosophy* 51: 483–499.
Hindriks, F. (2022): Institutions and their strength, *Economics and Philosophy* 38: 354–371.
Hindriks, F. & Guala, F. (2021): The functions of institutions: Etiology and teleology, *Synthese* 198: 2027–2043.
Hindriks, F. (2025): *The Structure of the Open Society: Social Ontology Meets Collective Ethics*, New York: Oxford University Press.
Hindriks, F. (n.d.): Functions in social institutions, in: Šustar, P. & Brzović, Z. (Eds): *Philosophy of Biological Functions*, London: Routledge.
Hindriks, F. & Guala, F. (2015): Institutions, rules, and equilibria: A unified theory, *Journal of Institutional Economics* 11: 459–480.
Jackson, F. & Pettit, P. (1992): Structural explanation in social theory, in: Charles, D. & Lennon, K. (Eds): *Reduction, Explanation, and Realism*, New York: Oxford University Press, pp. 97–131.
Lewens, T. (2015): *Cultural Evolution*, New York: Oxford University Press.
Lewens, T. & Buskell, A. (2023): Cultural evolution, in: Zalta, E.N. & Nodelman, U. (Eds): *Stanford Encyclopedia of Philosophy* (Summer 2023 Edition), https://plato.stanford.edu/archives/sum2023/entries/evolution-cultural/
Lewis, D.K. (1969): *Convention: A Philosophical Study*, Cambridge, MA: Harvard University Press.
Little, D. (1991): *Varieties of Social Explanation: An Introduction to the Philosophy of the Social Sciences*, Boulder: Westview Press.
Mesoudi, A. (2011): *Cultural Evolution, How Darwinian Theory Can Explain Human Culture and Synthesize the Social Sciences*, Chicago: University of Chicago Press.
Millikan, R.G. (1999): Historical kinds and the 'special sciences', *Philosophical Studies* 95: 45–65.
Nichols, R., Charbonneau, M., Chellappoo, A., Davis, T., Haidle, M., Kimbrough, E.O., Moll, H., Moore, R., Scott-Phillips, T., Purzycki, B.G. & Segovia-Martin, J. (2024): Cultural evolution: A review of theoretical challenges, *Evolutionary Human Sciences* 6: e12.
Pettit, P. (2007): Resilience as the explanandum of social theory, in: Bedi, S. & Shapiro, I. (Eds): *Political Contingency: Studying the Unexpected, the Accidental, and the Unforeseen*, New York: New York University Press, pp. 79–96.
Reydon, T.A.C. (2023): The proper role of history in evolutionary explanations, *Noûs* 57: 162–187.
Richerson, P.J. & Boyd, R. (2005): *Not by Genes Alone: How Culture Transformed Human Evolution*, Chicago: University of Chicago Press.
Richerson, P.J. & Henrich, J. (2012): Tribal social instincts and the cultural evolution of institutions to solve collective action problems, *Cliodynamics* 3.
Satz, D. & Ferejohn, J. (1994): Rational choice and social theory, *Journal of Philosophy* 91: 71–87.
Schotter, A. (1981): *The Economic Theory of Social Institutions*, Oxford: Oxford University Press.
Schulz, A.W. (2021): What's the point? A presentist social functionalist account of institutional purpose, *Philosophy of the Social Sciences* 52: 53–80.
Searle, J.R. (1995): *The Construction of Social Reality*, New York: The Free Press.
Smith, M. (1982): *Evolution and the Theory of Games*, Cambridge: Cambridge University Press.
Tuomela, R. (2002): *The Philosophy of Social Practices*, Cambridge: Cambridge University Press.

Ullmann-Margalit, E. (1977): *The Emergence of Norms*, Oxford: Clarendon Press.
Valković, M. (2023): Ontological and methodological limitations of certain cultural evolution approaches, *Philosophy of the Social Sciences* 54: 279–301.
Vermaas, P.E. & Houkes, W. (2003): Ascribing functions to artefacts: A challenge to etiological accounts of functions, *The British Journal for the Philosophy of Science* 54: 261–289.
Walsh, D.M. & Ariew, A. (2013): A taxonomy of functions, *Canadian Journal of Philosophy* 26: 493–514.
Watkins, J.W.N. (1959): The two theses of methodological individualism, *The British Journal for the Philosophy of Science* 9: 319–320.
Wouters, A.G. (2003): Four notions of biological function, *Studies in History and Philosophy of Science Part C: Studies in History and Philosophy of Biological and Biomedical Sciences* 34: 633–668.
Wright, L. (1973): Functions, *The Philosophical Review* 82: 139–168.
Young, P.H. (1998): *Individual Strategy and Social Structure: An Evolutionary Theory of Institutions*, Princeton: Princeton University Press.
Young, P.H. (2015): The evolution of social norms, *Annual Review of Economics* 7: 359–387.

Part II

Power

5 Can Cultural Evolution Accommodate Power? Race, Gender, and Lewontin's Critique

Azita Chellappoo

5.1 Introduction

Power in its various forms is a central feature of society. Power can be defined in many ways, both in terms of the individual and in terms of groups or structures. Power can be the capacity of one individual to impose their will on another, to affect others' actions through mechanisms of punishment, to shape the limits of political agendas, or to influence others' preferences (Dahl, 1957; Bowles & Gintis, 1992; Bachrach & Baratz, 1962; Lukes, 1974). Power can also be positive, in the sense that it makes new actions possible (Burman, 2023). Power can be understood as a property of groups, a feature of structures, or diffuse within networks of social relations (Arendt, 1970; Giddens, 1984; Foucault, 1982). These are just some of the options; there are, of course, many alternative accounts.

What exactly power is remains contested. However, what is clear is that power is a fundamental part of the social and cultural world. Any approach that aims to provide an overarching framework for understanding and explaining culture must take account of power. Therefore, the claim that cultural evolution cannot do so (advanced by Richard Lewontin, among others) is a serious challenge. For Lewontin, it is cultural evolution's commitment to methodological individualism that renders it unable to sufficiently deal with power, leaving its explanations of cultural change seriously inadequate (e.g., Fracchia & Lewontin, 2005).

Within social ontology, challenges have been mounted against ontological individualism, thereby casting doubt on the ability of individual-level theories to provide complete explanations of social phenomena (Epstein, 2009, 2015; Haslanger, 2020; Little, 2024). Possible higher-level entities that might populate our social ontologies include structures or social groups. In this chapter, I explore the extent to which Lewontin's diagnosis of the limits of cultural evolution's individualism still stands, through consideration of how cultural evolution might deal with racial and gender hierarchies.[1]

DOI: 10.4324/9781003643845-7

I begin by outlining Lewontin's critique and possible responses to the problem (reduction, restriction, and presumption, as identified by Lewens (2015)) in Section 5.2. In Section 5.3, I introduce some ways in which various cultural evolutionary approaches handle social groups and power in individualistic terms. I suggest that race and gender hierarchies are examples of structural power relations that at least some cultural evolutionary frameworks struggle to accommodate, as Lewontin's critique would predict. I suggest that the strategies of reduction (reducing these structural phenomena to the aggregated effects of individual interactions), restriction (restricting the application of cultural evolutionary frameworks to domains where higher-level phenomena do not have meaningful effects), presumption (modeling higher-level phenomena as fixed background conditions), and a cultural niche construction approach (Baraghith and Jordan, Chapter 6) do not provide satisfactory resolutions.[2] However, despite the apparent problem that structural phenomena such as social hierarchies pose, there is a growing body of work that uses evolutionary game theory models to generate insights into the conditions under which racial and gender hierarchy can emerge. In Section 5.4, I outline this work and its theoretical commitments. I then assess the extent to which the success of these models challenges Lewontin's critique in Section 5.5. In Section 5.6, I conclude and sketch some possible connections between research in social ontology and cultural evolution that come out of this analysis.

5.2 Individualism and Lewontin's critique

The inability to account for social power was a central component of Lewontin's skepticism about the fruitfulness of evolutionary approaches to culture. For Fracchia & Lewontin (2005), this shortcoming is a consequence of the ways in which Darwinian theories of cultural change reduce society to "a collection of atomistic individuals" (69). This methodological individualism is a necessary part of approaches that understand cultural change in terms of the basic components of Darwinian evolution, taking as their starting point a population of individuals exhibiting random variation, heredity, and differential reproduction. Not all cultural evolutionary work is selectionist (aiming to characterize cultural change in terms of processes analogous to natural selection), but much of it is nevertheless fundamentally committed to "population thinking" (Richerson & Boyd, 2005). Population thinking, a term coined by Ernst Mayr (1959), can be understood as "the kind of thinking one engages in when one explains the behavior of a unit composed of varied parts in terms of the properties of those parts and their interactions" (Lewens, 2009: 249).

Population thinking exemplifies a commitment to both ontological and methodological or explanatory individualism: the social world is made up of nothing more than individuals, and the best explanations are those that are based on individuals and their interactions.[3] Furthermore, population thinking explanations do not attend to the details of specific individuals. Rather, individuals within a population are idealized as "relatively autonomous items of different types" (Claidière et al., 2014: 2), with cultural change being "the frequency of types changing over time" (Claidière et al., 2014: 2). Accordingly, Tim Lewens (2015) describes cultural evolutionary approaches that adopt population thinking as "kinetic theories", akin to the kinetic theory of gas, which understands features of gases in terms of the aggregated effects of interactions of individual molecules with each other and the background environment.

For Lewontin, reducing societies to populations (collections of individuals) in this way leaves out phenomena of power:

> However accomplished, the dissolution of societies into populations or, as in more nuanced approaches, the reduction of differential social power to the status of a subordinate variable, precludes the possibility that social systems might have properties unique to them as organized systems, that is, that social relations might be characterized by structures of unequal power that affect individual social behavior and the fitness of cultural traits. This dissolution means, in turn, that social hierarchy and inequality are explained as just the consequence of the differential cultural fitness of individuals or of the cultural traits they bear, rather than, say, as a consequence of antagonistic and exploitative social relations.
>
> (Fracchia & Lewontin, 2005: 70)

As hierarchy, inequality, and various mechanisms of centralized control are key features of societies, any approach to culture that struggles to take them into account is fundamentally flawed. One reason why power presents such a challenge for evolutionary or populational approaches is that power can provide a competing explanation for cultural change:

> the success or "selectability" of a solution, whether it will be short- or long-lived, depends not just on its "fitness" in regard to specific problems, but also and crucially on the amount of power behind it – power that can force its "selection", at least temporarily, regardless of whether it is a "fit" solution.
>
> (Fracchia & Lewontin, 2005: 21–22)

For example, the spread of one religion over another might not be explainable in terms of its functional benefits or individual learning biases, but rather through channels of top-down control, such as a powerful ruler, government, or set of institutions (Lewontin, 2005).

Lewontin's critique has since been taken up and expanded upon (see also Baraghith and Jordan, Chapter 5, for additional discussion). In the face of worries that cultural evolutionary theories struggle to account for powerful individuals and organizational or institutional power, Lewens (2015) suggests three possible strategies for cultural evolutionists: restriction (limiting the application of cultural evolutionary models), presumption (modeling organizations or institutions as fixed background conditions that affect individual action, but are not themselves explananda), and reductionism (understanding complex organizations or institutions themselves as reducible to individual-level populational models).

None of these strategies provides a clear resolution to Lewontin's critique. Both restriction and presumption place significant limits on what cultural evolution can explain. Given that societies are full of various power relations, hierarchies, and inequalities, theories that either restrict themselves to cases where power is not a crucial factor in cultural change, or treat power wielded by organizations and institutions as mere background conditions without considering dynamic relationships between individuals and these institutions, will be extremely limited in their application, particularly when aiming to explain features of contemporary societies. This is in tension with the lofty aims of cultural evolutionary approaches: much work in the field aims to understand and explain complex phenomena such as technological innovation and development, cultural transmission through digital media, and institutional dynamics (Henrich, 2001; Acerbi, 2016; Richerson & Henrich, 2012). Furthermore, some of the promise of cultural evolution lies in its potential to inform solutions to pressing global issues, such as in environmental governance and sustainability (Brooks et al., 2018).

If restrictions or presumptions are unpalatable, will taking the reductionist strategy work? This response is an affirmation of cultural evolution's individualism, committing to explaining complex group structures or institutional dynamics through aggregated effects of individuals and their interactions. Baraghith and Jordan (Chapter 6 in this volume) suggest one for employing the reductionist strategy to accommodate power, through drawing on an analogy with r- and K-selection. Another group of researchers has adopted the reductionist strategy in a different way, turning to evolutionary game theory to explain phenomena of power (O'Connor, 2019; Bright et al., 2025). I will return to an assessment of the reductionist strategy in the following section, but for now, I will note that it remains controversial, analogous to persistent controversies over individualism in the social sciences (Kroeber, 1917; Lewontin & Levins, 2007; Sawyer, 2005; Valković, 2023).

5.3 Power, Race, and Gender

If Lewontin's critique is broadly correct, then it seems that cultural evolutionary theories will struggle not only to accommodate the influence of powerful individuals or organizations but also to account for or understand social groups that are themselves characterized by power inequalities, such as race or gender. To explore the limits of cultural evolution's individualism, in this section, I first outline some of the ways in which cultural evolutionary work handles social groups and power. I then contrast this with race and gender as examples of social groups characterized by power hierarchies.

Cultural evolution's individualist ontology is reflected in, for example, understandings of culture itself as:

> information capable of affecting individuals' behavior that they acquire from other members of their species through teaching, imitation, and other forms of social learning.
>
> (Richerson & Boyd, 2005: 5)

Here, culture itself is defined on the level of the individual. For Richerson & Boyd (2005), culture is mostly something that resides in the brains of individuals. Advocates of cultural epidemiology, a competing school of cultural evolution to selection-based approaches, share a similar view of culture as information transmitted between individuals in a range of ways, including through mental representations and behavior (Sperber & Claidière, 2008).

Nonetheless, ontological and explanatory individualism does not preclude attending to supra-individual phenomena. In fact, much cultural evolutionary work deals with social groups, including work that focuses on cultural group selection (processes analogous to natural selection operating at the level of groups). However, these groups are themselves understood in terms of the aggregated effects of small-scale individual interactions. For example, one process of cultural group selection is selective imitation, where individuals preferentially copy the behaviors of other individuals from more successful groups, leading to the spread of group-beneficial cultural traits (Richerson et al., 2016). Another possible path lies in work on cultural niche construction, which offers resources to represent social structures in terms of network relations that structure the interactions of individuals (Baraghith and Jordan, Chapter 5).

Cultural evolutionary work has also attended to phenomena of power to some extent, although again within the constraints of individualist population thinking. In accounts of learning biases that are central to some cultural evolution approaches, certain individuals will have more power to influence other individuals within a population (where influence means the likelihood of cultural traits being copied). For example, prestige bias is a bias to copy the most "prestigious" individuals, and therefore, prestigious individuals will have power in the sense that their behaviors will be more likely to spread through the population (Jiménez & Mesoudi, 2019).

The concept of prestige itself is often understood in individualistic terms: prestigious individuals are the ones whom learners are seeking to copy (Henrich & Gil-White, 2001; Hindriks, Chapter 4; Baraghith and Jordan, Chapter 6).

The attempted application of cultural evolution to pressing global challenges, such as sustainability, has spurred further consideration of power within institutions. Safarzyńska & van den Bergh (2010) sketch a framework of power in the context of using cultural group selection to explain environmental institutions and policies, characterizing power in the following terms:

> Power of an individual can be measured in terms of wealth, access to powerful others, or a level in a hierarchy of decision-making (legitimised source of power). In addition, power of a group can be quantified by its size, frequency of cooperators, or total cumulative wealth of its members.
>
> (Safarzyńska & van den Bergh, 2010: 750)

Here, again, power is conceived of in mostly individualistic terms.[4]

What are the limits of cultural evolutionary ways of understanding and accommodating power? I suggest that, expanding on Lewontin's critique, social groups that are defined in terms of structural power relations appear to present a challenge to cultural evolution's individualism.

I offer race and gender as two examples of cultural phenomena that are often understood in structural or systematic terms in the social sciences and in social ontology. I also choose these as they are central aspects of many societies, and so presumably have a meaningful influence on trajectories of cultural change, and therefore are relevant to the phenomena cultural evolutionary research aims to explain. Although not without challenge, I take as a starting point Haslanger's (2000) well-known account of race and gender, structurally defined in terms of power relations or social hierarchies. According to Haslanger:

> A group is racialized iff_{df} its members are socially positioned as subordinate or privileged along some dimension (economic, political, legal, social, etc.), and the group is "marked" as a target for this treatment by observed or imagined bodily features presumed to be evidence of ancestral links to a certain geographical region.
>
> (Haslanger, 2000: 44)

If we accept this account, then the racial category "Black" or the gender category "woman" are not just groups that are contingently oppressed or discriminated against, but are themselves "social groups defined within a structure of social relations" (50). Social structures are networks of social relations that shape possibilities and constraints for individual actors, in terms of their own practices, their resources, and their interactions with other individuals (Haslanger, 2023).

How might a cultural evolutionary approach focused on individuals handle these sorts of power hierarchies produced by social structures? As outlined in the previous section, possible responses are reduction, restriction, and presumption. I now consider each of these in turn, and lastly explore the pursuit of cultural niche construction to accommodate social structures, which arguably does not fit neatly into any of these three strategies.

Is a reduction possible? Can racial or gender hierarchies be reduced to individual-level phenomena? Critics of ontological individualism argue that social facts do not reduce to (or, for weaker versions of ontological individualism, do not supervene on) individual properties (Epstein, 2009, 2015; Haslanger, 2020). This is because "constituent material parts of entities also matter in determining their social features, so the social features do not depend wholly on the attitudes and actions of individuals, either constitutively or causally" (Haslanger, 2020: 519). For example, a corporation going from financially solvent to insolvent could happen without any individual awareness or action (say, from a midnight power surge that destroys crucial material infrastructure) (Epstein, 2015). In this example, it is not any individual, and plausibly not any individualistic property – no matter how broadly construed – that explains the transition to insolvency.[5]

An equivalent point can be raised in the context of racial or gender hierarchy. Consider the following case:

> *Real Estate*. For centuries, members of R1 systematically targeted the economic resources of members of R2 for (unjust) suppression, first through a state-supported system of slavery, then through a state-supported system of segregation. Eventually, there was a racial rapprochement, where all state-supported means of R2 's economic repression (including criminal laws, voting rights, and tax-and-transfer schemes) were abolished. However, the rapprochement was not complete in the sense that no reparations were made to members of R2, and therefore the postrapprochement era begins with R1s uniformly having more wealth than R2s. Two hundred years later, there are no longer any attitudes, behaviors, or beliefs, among either R1s or R2s, which we would ordinarily identify as racist. However, there also is no change in mortgage policies, which ensures that more, and more beneficial, mortgages go to those with better credit ratings, and better credit ratings are assigned to those with greater assets. As a direct result of this state

> of affairs, R2s still have fewer good homes, less wealth, and a weaker sense of economic security than R1s have.
>
> (Glasgow, 2009: 72–73)

Joshua Glasgow uses this hypothetical case to argue for antireductionism about institutional racism (that racism can "bottom out" at the institutional level, without the presence of any racism at the individual level). In the above example, there is (arguably) structural or institutional racism. It therefore seems that race (or racial hierarchy) still exists as a social entity. In this case, it is not facts about or properties of individuals that perpetuate the social hierarchy. Fixing the facts about individuals would not determine higher-level facts because the higher-level facts depend in part on properties or entities that are not plausibly individual (e.g., bank mortgage policies).

If reduction is problematic, might restriction be more appealing? Racial and gender hierarchies are examples of power inequalities that are likely to have effects on patterns of cultural change in many domains. This means that a restriction to cases where individual-level explanations (reflective of underlying ontological commitments) suffice may be a serious constraint on cultural evolution's domain of application. For example, cultural evolution has been proposed as a powerful approach to understanding cultural transmission across digital media (Acerbi, 2016, 2019). However, racial hierarchies (for instance) are part of the network of social relations that structure individual interactions (Daniels, 2012). This has implications for transmission patterns generated by learning biases that are commonly part of cultural evolutionary analysis. For example, the patterns generated by prestige bias in digital contexts will depend on who is identified as prestigious. Prestige is related to power, and racial inequalities can shape the prestige or influence of individuals online in various ways (Messias et al., 2017; Freelon et al., 2022). Another proposed learning bias is self-similarity bias: a bias toward copying individuals similar to ourselves in certain respects, with commonly cited features being sex, age, or ethnicity (Henrich, 2004; Henrich & Broesch, 2011). Who we perceive as relevantly similar to ourselves will, of course, be affected by the racial schema operative in a given context, as part of what racial ideologies do is to produce (and inferiorize) difference (Omi & Winant, 1986).

Given the issues with reduction and restriction, perhaps presumption (taking race and gender hierarchies to be fixed background conditions or constraints within which individuals interact) seems to be the most promising option. However, this is only feasible if there is no (or no significant) effect of individuals and their interactions on the background structures themselves. This is not a plausible assumption: racial and gender structures are in dynamic interaction with individuals and material technologies, both shaped and shaping (Kang, 2000; Parker & Song, 2006; Frey et al., 2022; Van Doorn, 2011; Marwick, 2013).

Finally, recent work in terms of cultural niche construction and network relations, exemplified by Baraghith's and Jordan's account in Chapter 6 of this volume, provides another possible way forward. In the authors' model, which focuses on structural dominance or coercion, individuals are represented as nodes in a network, whose interactions are structured in particular ways. Power is represented in terms of the number of individuals (nodes within the network) that any one individual can compel to alter the behavior of (translated into a weighting assigned to each interaction between nodes). Individual positions within this weighted network of interactions could be described as their cultural niche.

This approach is, I think, not easily characterized in terms of the three strategies above. In their initial model, the structure of the interactions and their weightings is presumed, although there is no reason to think that this approach, in general, would require this – as in biological niche construction models, the development of the cultural niche itself could be modeled. It is a direct attempt to understand power structures in cultural evolutionary terms, rather than restricting cultural evolution to non-structural phenomena. And, the authors suggest that it is not a reduction, as the dynamics of the network are not reducible to the individuals themselves (individuals can be replaced at each node, while the structure of the relevant interactions remains the same). On the other hand, one could argue that this is reductive in the sense that the model remains focused on individual interactions, rather than incorporating higher-level entities.

Could this approach accommodate racial and gender hierarchies? I remain skeptical. To be clear, Baraghith's and Jordan's model is an initial step, which focuses on dominance and coercion in particular. Therefore, any inability of the current model to illuminate racial or gender hierarchies does not mean that further development of the general approach would necessarily have this limitation. Nevertheless, I will briefly consider how their model might grapple with Glasgow's *Real Estate* case. As a reminder, in Glasgow's case, historical racial oppression has resulted in a situation where members of the race R2 have fewer economic resources than members of the race R1. The mechanisms of producing this inequality are no longer present, although there were also no reparations made. Presently, no individuals display racist attitudes, behaviors, or beliefs. Mortgage policies mean that more and more beneficial mortgages go to those with better credit ratings (those with more assets). This means that R2s still have fewer good homes, and less wealth and economic security compared to R1s.

How would this situation be modeled in terms of weighted interactions between nodes in a network? We might imagine a network that includes nodes for the heads of banks, and various other employees involved in determining who is offered a mortgage and on what terms, and the individuals seeking mortgages. Presumably, the higher-up employees in the bank would have greater power (in Baraghith's and Jordan's sense), as they would be able to compel lower-level employees to change their behavior (e.g., by changing who they give out mortgages to). And, lower-level employees have power over mortgage-seekers, as they can change behavior through offering or denying mortgages. Does this speak to the nature of the racial hierarchy here? I suggest not. The hierarchy I have sketched would remain the same whether or not there was racial inequality in mortgage policies – in a society where there were no races, but mortgage policies still go to those with better credit ratings, the power structure suggested here would remain the same.

Perhaps this is not the relevant network. Another option is to model (or to add to the previous model) a network where nodes occupied by R1s generally have more power than nodes occupied by R2s. However, if understanding power in terms of the ability to compel behavior, this might not quite capture the relevant inequality in Glasgow's case.[6] One could imagine that there is no clear distinction between R1s and R2s in their capacity to compel the behavior of others. For example, R2s and R1s might be equally represented in government and corporations. They might even have, on average, equal incomes. Yet, because of the history of racial oppression, R2s still have fewer assets than R1s (having inherited fewer assets from their parents, grandparents, and so on). Furthermore, R2s might have greater capacity to compel the behavior of others – imagine a scenario where R2s disproportionately take on managerial or government roles, while R1s disproportionately pursue creative or freelancing careers, or a scenario where R2s disproportionately pursue dominance strategies while R1s pursue prestige strategies. These scenarios could occur while the inequality in mortgage policies and its economic consequences persist. On a broad reading of what it means to be able to compel others' behavior, wealth in itself could be a form of power insofar as it is a way to compel behavior. Greater wealth might mean being able to compel a greater number of individuals to, say, provide goods and services. However, imagine a scenario where R1s have more assets than R2s, but they prefer to save rather than spend their wealth – rather, their greater wealth gives them a comforting feeling of economic security, contributing to a racial inequality in psychological wellbeing. The point is just that, in the *Real Estate* case, modeling the structural inequality as a network of weighted interactions between individuals, where individuals occupy their own cultural niche, does not seem to offer much in the way of explanation or insight into what is really going on.[7] It fails to satisfactorily capture the role of higher-level institutional facts in generating inequality (and, under a Haslangerian definition, generating race itself).

I have suggested that structural racial and gender hierarchies present a challenge to cultural evolution's individualism, providing further support to Lewontin's critique. This is perhaps reflected in the lack of attention to race and gender in cultural evolution work on contemporary issues (e.g., digital media or sustainability). However, in recent years, there has been a growing body of cultural evolution work that targets inequality directly, seeking to understand and explain the emergence and persistence of categories characterized by differential power relations such as race and gender. In the next section, I outline what this research shows. I explore the extent to which it might challenge a version of Lewontin's critique in Section 5.5.

5.4 Cultural Evolutionary Models of Inequality

Historically, race or gender (understood as structural power relations) have not been primary concerns for cultural evolutionary work. However, this is no longer the case: a growing body of recent research has used cultural evolutionary approaches to illuminate the development of racial and gender categories as sites of inequality. In this section, I provide an overview of this research and highlight the ways in which it uses population thinking to model systemic or structural phenomena, before assessing the extent to which it challenges Lewontin's critique in Section 5.5.

A prime example of this kind of work comes from Cailin O'Connor's 2019 book *The Origins of Unfairness: Social Categories and Cultural Evolution*. O'Connor uses game theory and evolutionary game theory to show how "inequity is the expected outcome of basic cultural evolutionary processes". In O'Connor's analysis, gender categories with associated norms and roles (e.g., gendered division of labor) provide solutions to coordination problems. In "complementary coordination games", actors must choose complementary actions in order to receive the highest overall payoff (e.g., when ballroom dancing, one person must lead and the other follow). In populations, social categories (like gender) can provide a solution to complementary coordination problems through identifying what role each actor should play. Gender is useful in performing this function because sex and gender markers are salient to us and easily recognizable. While social categories are beneficial in terms of facilitating coordination, they also confine people to particular strategies or social roles, and in doing so open the door to inequality. O'Connor uses evolutionary game theory to show that social categories will evolve to facilitate complementary coordination, and that once these types are in place, inequality robustly emerges under minimal conditions. This can generate inequalities in the division of labor and distribution of resources.

As with other cultural evolutionary approaches, O'Connor's analysis is on the level of the individual. For example, the concept of "power" is operationalized in terms of individual interactions: a group holds power over another group when members of that group have the potential to lower the payoffs of members of the other group. From starting with populations of individuals and aggregating small-scale individual interactions, O'Connor provides possible explanations for the emergence and re-emergence of inequitable conventions: a picture appears of systemic power differentials. Her account also provides possibilities for remedying social inequality, indicating that just addressing psychological bias is not enough, that inequality will reliably re-emerge without constant vigilance, and suggesting the importance of intervening on payoffs or ability for type conditioning to occur.

Similarly, Bright et al. (2025) use models from evolutionary game theory to analyze theories of racial capitalism (theories that give race a functional role in the survival and spreading of systems of capitalist production). As in O'Connor (2019), power is operationalized as higher disagreement points in bargaining games: if bargaining fails, powerful social groups are in a better position compared to less powerful ones. Therefore, more powerful social groups can better afford to let bargaining fail, and this translates into a bargaining advantage (less powerful social groups are incentivized to accept unequal bargaining outcomes as their payoffs will be worse if bargaining fails entirely). Game theoretic models demonstrate that race has functional advantages in terms of its use as a social marker that grounds inequality, due to its relative inflexibility (so members of the disadvantaged group cannot easily switch out of that group), its reliability (ease of identifying group members), and its heritability. Additionally, cultural evolutionary models have been used to model the emergence of intersectional inequality, where structures of oppression like race and gender intersect to produce particular disadvantage (O'Connor et al., 2019).

This work all shares a commitment to a population thinking approach: societies are modeled as populations of idealized individuals whose small-scale interactions produce overall effects. Power is understood in terms of these individual interactions. One criticism of the work of prominent cultural evolution researchers such as Alex Mesoudi has been that their ontologies are absent of any structural or systematic phenomena (Valković, 2023). It has been suggested that these underlying ontological commitments could shape methodological choices, leading to a neglect of or inability to account for such structural phenomena in explanations. In contrast, the work of O'Connor, Bright, and colleagues starts with an individualist ontology precisely to account for features of systems or structures. They generate models or explanations of system-wide inequality and aspects of structures (e.g., types that position individuals as relatively disadvantaged or advantaged).

5.5 Revisiting Lewontin's Critique

Does the success of this line of research present a challenge to Lewontin's critique? The answer is both yes and no.

The headline-level claim of Lewontin's critique is that cultural evolutionary approaches cannot accommodate power. The research of O'Connor, Bright, and others demonstrates this not to be true. Lewontin's diagnosis, that it is the reduction of societies to populations that leaves cultural evolution unable to take power into account, is also in trouble. It is precisely by modeling societies as populations that evolutionary game theoretic approaches can generate explanations of unequal or oppressive power relations. The dynamics of individual-level bargaining games can explain why systems can arise where certain groups have more power than others – there can be stable equilibria where individuals are incentivized to accept unfair distributions because rejecting them leaves those individuals worse off. Members of certain groups, therefore, have the power to, for example, spread particular cultural traits at the expense of others – some individuals or groups have the ability to force the selection of cultural traits, whether or not they are fit solutions.

However, to leave it at this would be unfair. In assessing Lewontin's critique, it should be recognized that it, while pointing to the trouble with population thinking, was primarily aimed at certain sorts of selection-based approaches. For example, Fracchia & Lewontin (2005) characterize the selectionist paradigm as saying that:

> "societies" and "states" have no systemic logic, that they consist only of "populations" of individual units whose relation to one another is the merely external one of competitive selection; the fitness of a practice or meme is a function solely of its own attributes and has no direct relation to the fitness of its competitors.
>
> (Fracchia & Lewontin, 2005: 21)

This may remain an apt critique of certain sorts of cultural selection approaches. If we want to explain the spread of a catchy song, it probably would not do to look solely at either content (features of the song itself that make it appealing to humans) or model biases (such as prestige or conformist bias). Factors such as whether a major record company is promoting it or not will be crucial in explaining why one song has many more listeners than another. However, in the evolutionary game theoretic models of racial and gender hierarchy under discussion, it is not the case that societies have no "systemic logic" or that there is no relation between individuals. It is precisely the relationship between individuals (e.g., relative differences in disagreement points) that generates the insights.

In some ways, Fracchia and Lewontin provide an accurate diagnosis of the problem with some kinds of selectionist approaches to culture, while perhaps being mistaken in exactly why this problem arises. They write:

> Because of its inescapable methodological individualism and the methodological randomness it spawns, the selectionist paradigm produces a massive blind spot toward the logic of social relations and socially-structured antagonisms; it reduces the social conflict produced by those antagonisms to the indifferent difference of competitive selection; and it fails to realize that the greater "success" or "fitness" of certain memes/practices is not a mere matter of random variation. The survival or "selection" of unfit practices is a necessary part of the social relations of inequality and of the social reproduction of the "fittest". This is not just an idiosyncrasy of, a "disanalogy" between, social and cultural as opposed to biological evolution. The active reproduction of "unfit" practices and memes is rather constitutive of the entire process of social reproduction.
>
> (Fracchia & Lewontin, 2005: 23)

Fracchia and Lewontin are correct that social antagonisms (or power relations) are a crucial part of understanding the spread of "unfit" practices, and that this is often not a part of the selectionist picture. However, in the work of O'Connor, Bright, and colleagues, the success or fitness of particular strategies is not reduced to a "mere matter of random variation", and is, in fact, part of the "social relations of inequality". It is precisely because of the dynamics that they identify that "unfit" practices can spread and persist (very generally because stable equilibria can be reached where moving away from this toward a more equitable situation will result in lower payoffs, at least in the short term). As an example of "unfit" social relations, Fracchia and Lewontin write that "As Marx so clearly explained, capitalism reproduces the "unfit" practice of wage labor" (2005: 23). Bright et al. (2025) provide a demonstration of how race can stabilize these inequitable capitalist systems:

> Race creates an asymmetry where those associated with a less advantaged social position receive their best payoffs by complying with the system. Those in advantaged positions can profit from this fact.
>
> (Bright et al., 2025: 9)

The insights generated from these models are not in direct contradiction with Fracchia's and Lewontin's pointing out of the gaps in cultural selection approaches at the time. However, what they perhaps show is that methodological individualism is less limiting than Fracchia and Lewontin suggest.

While the work of O'Connor, Bright, and colleagues takes an individual-level approach to explore certain aspects of hierarchical systems, this does not necessarily justify an individualist ontology.[8] Firstly, the explanations provided by these models are only part of the story (as they are intended to be). The evolutionary game theoretic models produced in this work are primarily how-possibly explanations for the emergence and stability of forms of inequality or hierarchy. They show that inequality robustly emerges given a certain set of minimal conditions, not a precise story for how inequality in fact emerged or a complete account of the various ways in which it is sustained.[9] They are not completely functional or causal explanations for racial or gender oppression. For example, these models provide possible explanations for the emergence of inequality from individual-level interactions, which are compatible with historical accidents or shocks (perhaps as a consequence of the idiosyncratic actions of a small number of individuals) that alter what actually occurs. However, the models show the existence of basins of attraction that will pull populations toward one equilibria or another, and so can say something about overall trajectories even in the face of particular disruptions. Additionally, these models are compatible with the subsequent generation of higher-level entities or structures that play a role in the maintenance or further development of the hierarchical structure of the system. Once inequality emerges from unequal bargaining conventions, members of the group with more power can shape structures, organizations, and institutions in ways that cement that inequality. These institutions can then perpetuate inequality even if there are changes on the individual level.[10] So, insofar as Lewontin's critique is a response to the claim that evolutionary approaches to culture can replace social-scientific approaches as complete accounts of society and culture, it still has force.

It should be noted that the level of emphasis placed on higher-level structures, while still compatible with the findings of these game theoretic models, may affect the usefulness of the prescriptions for intervention. If non-individualistic facts about, or properties of, for example, institutions are key drivers of inequality, then perhaps altering individual payoffs or even altering the ability for people to type-condition (choose their strategy based on the other actor's type) would not be sufficient to dismantle hierarchy or inequality entirely[11] (e.g., the inequality arising from mortgage policies in Glasgow's case in Section 5.3). That said, intervening on the incentive structure for individuals may impact the ability to make changes to these institutions.

Furthermore, in the work of O'Connor, Bright, and colleagues, the details of learning biases and cultural transmission are largely set aside. Learning biases are relevant for how individuals adopt conventions and recognize types, but the particulars of these are not the primary concern, although, for example, imitation of the successful plays a role in some models (Bright et al., 2025). However, a fuller understanding of the reproduction of power relations will depend on the particulars of cultural transmission (how conventions and strategies are adopted and spread). Additionally, filling in the how-possibly explanations with how-plausibly explanations will involve investigation of, for example, the details of the payoff structures and the ways in which individual incentives are shaped or constrained. A natural way of situating these models within a broader understanding of cultural evolution might be in terms of an "evo-devo" approach to culture, as proposed by William Wimsatt and James Griesemer. This approach rejects methodological individualism, claiming that:

> The socializing trajectories of individual learning and development are unintelligible without the structured shaping, guidance, and reinforcement that social institutions and organizations provide. These are not just environments for learning, but interactive products of culture – both produced by and feeding back to form enculturated individuals. Adequate theory requires an integral account of both levels and their reciprocal interaction.
>
> (Wimsatt & Griesemer, 2007: 228)

An "evo-devo" approach takes cultural development into account, placing an emphasis on the idea of "scaffolding": structures or relations that provide ways for cultural traits, skills, or strategies to be constructed or acquired (Wimsatt & Griesemer, 2007; Wimsatt, 2013; Love & Wimsatt, 2019). Scaffolds provide supports that enable certain developmental trajectories. One way of understanding racial and gender hierarchies is in terms of their function as scaffolds, reproducing particular population structures and cultural trajectories. Wimsatt & Griesemer (2007) discuss the role of scaffolding in gender differences in medical training (resulting in asymmetries in income, career development, and prestige), and racial differences in access to high-quality education.

In sum, Lewontin's critique appears weakened by the ability of evolutionary game theoretic models to generate insights into structural inequality. However, it remains powerful in its appraisal of certain forms of cultural evolutionary work or certain sets of assumptions, and in its reminder that social power relations are central to the project of understanding culture. In the final section, I provide some brief reflections on the ways in which the above discussions suggest possibilities for mutually beneficial engagement between social ontology and cultural evolution.

5.6 Conclusions and Connections

In returning to Lewontin's critique of cultural evolution, I have suggested that it still has force in important ways. I have drawn on critical discussions within social ontology to argue that structural racial and gender hierarchies present a challenge to cultural evolution's individualism, providing further support to the claim that cultural evolution cannot adequately account for power relations. However, recent developments that make use of evolutionary game theory to model racial and gender inequality appear to challenge the claim that it is cultural evolution's individualist commitments that render it unable to accommodate power. I have suggested that this work does demonstrate that an individualist approach can generate more insights into structural phenomena than perhaps Lewontin anticipated, although these remain incomplete. Finally, I want to briefly explore what might come out of the above discussions of power and social hierarchy within cultural evolution in terms of making connections between research within social ontology and cultural evolution.

Firstly, debates within social ontology over ontological individualism can be informative for cultural evolutionary work. If critiques of ontological individualism are broadly correct, then this weakens the basis for the explanatory individualism at the heart of many cultural evolutionary approaches (and may instead provide some support for non-individualistic cultural evolution frameworks, such as "evo-devo" approaches). This is particularly relevant for cultural evolutionary ambitions to provide an overarching framework for the investigation of culture, or for cultural evolutionary research that aims to explain and generate interventions in areas where complex institutions and higher-level structures abound.

There is also a connection between current debates in social ontology and the emphasis within cultural evolution on variation and change. This chimes with recent work in social ontology that emphasizes process, such as Daniel Little's account of the social, where

> relationships between the social world and the particular constitution of human actors at a given time are wholly recursive: social arrangements at $\mathbf{t}_i$ causally influence individuals at $\mathbf{t}_i$; actions and transformations of individuals at $\mathbf{t}_{i+1}$ lead to change in social arrangements in $\mathbf{t}_{i+1}$; and so on. So neither social arrangements nor individual dispositions are temporally or causally prior; each is causally dependent upon the other at an earlier time period . . . , it implies that neither actors nor structures have "essential natures" or fixed and unchanging properties. Rather, the properties of social structures are influenced by the past and present actions and thoughts of actors and the prior characteristics of structures; while the mental characteristics of present actors are shaped by the ambient social arrangements within which they develop.
>
> (Little, 2024: 191–192)

The rejection of essentialism finds an echo in Ernst Mayr's initial characterization of Darwin's "population thinking" as a shift that overturned essentialist modes of thought in biology. Mayr's rejection of typological thinking gave centrality to the individual. However, non-individualistic cultural evolutionary frameworks, such as an "evo-devo" approach that emphasizes processes of scaffolding and generative entrenchment, combine an emphasis on ongoing change with reciprocal causation and complex interactions between entities and structures at various levels.

Additionally, cultural evolutionary work may usefully inform some debates within the social ontology of race and gender. For example, Phila Msimang outlines a minimal account of social race in the following way:

> The characteristics of a social race are that they are stereotyped groups (typically identifiable by their phenotypic features and assumed ancestry), the presumption that racial identity is fixed, and the belief that race is a genealogical and heritable kind of group belonging.
>
> (Msimang, 2019: 15)

This chimes with the insights from Bright et al. (2025), who highlight the functional role that race plays as a social tag: it is easy to identify, relatively inflexible, and heritable. Although showing that race has certain characteristics central to its functional role in a hierarchical system does not provide a solution to ontological debates around race, it can inform these discussions by giving indications of what features are particularly central to the concept of race and providing possible reasons for why race has these features.

These are only a few possibilities for fruitful connections between research programs within social ontology and cultural evolution: further cross-fertilization of these two approaches could generate many more insights into the nature and workings of the social world.

Notes

1 Lewontin, together with the historian Joseph Fracchia, proposed several challenges to cultural evolutionary frameworks, of which their inability to deal with power was only one (although prominent). Other issues include problems with adaptationism, demarcating units of culture, and specifying rules of transmission (Fracchia & Lewontin, 1999, 2005). However, in this chapter, I focus only on the aspect that concerns power, and in what follows I use the phrase "Lewontin's critique" to mean the claim that cultural evolution cannot satisfactorily account for power relations.

2 In Chapter 6 of this volume, Baraghith and Jordan similarly suggest that restriction and presumption have limitations, although they offer some hope for the reductionist strategy, based on an analogy with r- and K-selection. In this chapter, I focus on another attempt to pursue reduction that draws on evolutionary game theory.

3 Hindriks, in Chapter 4 of this volume, highlights the attempt from cultural evolution researchers such as Alex Mesoudi to restrict cultural evolution's individualism to the methodological while avoiding a commitment to ontological

individualism. However, Hindriks highlights the obstacles in drawing this distinction, and suggests that Mesoudi's stance is best interpreted as a commitment to both ontological and methodological individualism.

4 Lewontin's critique of inability of cultural evolutionary accounts to accommodate power, and the arguments offered in support in this section, will still stand regardless of whether power should be understood as a property of an individual, or if power itself can be structural. What matters is that power results or is expressed in structural (higher-level) features of a society, such that power relations cannot be explained without reference to higher-level entities. This can be the case even if those structural features give rise to power differentials as understood on an individual level.

5 For further discussion of this case, see Chapter 4 in this volume.

6 This issue might be somewhat remedied with an alternative, richer, account of power. This might include an account of power as positive or negative, or making use of the distinction between deontic and telic power (Burman, 2023: Chapter 7). This would rely upon the development of a useful operationalization of such an account.

7 I want to emphasize again that this does not show that understanding inequality in terms of network relations is necessarily flawed. As discussed in the following section, an approach that focuses on interactions between individuals can generate important insights about social hierarchies. However, it may be limited in the scope of its explanations, and may need supplementation with explanations in terms of higher-level entities.

8 As Hindriks (Chapter 1) suggests, methodological individualism could be replaced with methodological pluralism, where some aspects of social phenomena are explained in individualist terms, and others in more complex terms. A holist ontology is compatible with methodological pluralism, and therefore with the position that the individualist work of O'Connor, Bright, and colleagues is fruitful while not being the complete picture.

9 As, for example, O'Connor et al. (2019) acknowledge in their work on intersectional disadvantage, "our models do not display many of the characteristics that contribute to real-world inequity, such as psychological features, or society-wide features like segregated neighborhoods, police violence, and gendered pay inequalities. Removing these contextual details makes clear that we are not building in the intersectional effects, but rather showing how they can fall out of basic assumptions about group membership and strategic interaction" (37).

10 A similar supplementation of a model focused on individual interactions with higher-level entities may also be fruitful, or indeed necessary, for the cultural niche construction approach proposed by Baraghith and Jordan (Chapter 6).

11 This is not a criticism of this work, as they do not claim that changing the incentive structure for individuals is the complete or only solution (but is, perhaps, a necessary part of the solution).

References

Acerbi, A. (2016): A cultural evolution approach to digital media, *Frontiers in Human Neuroscience* 10: 636.

Acerbi, A. (2019): *Cultural Evolution in the Digital Age*, Oxford: Oxford University Press.

Arendt, H. (1970): *On Violence*, New York: Harcourt, Brace, Jovanovich.

Bachrach, P. & Baratz, M.S. (1962): Two faces of power, *The American Political Science Review* 56: 947–952.

Bowles, S. & Gintis, H. (1992): Power and wealth in a competitive capitalist economy, *Philosophy & Public Affairs* 21: 324–353.

Bright, L.K., Gabriel, N., O'Connor, C. & Taiwo, O. (2025): On the stability of racial capitalism, *Ergo an Open Access Journal of Philosophy* 12: 42.

Brooks, J.S., Waring, T.M., Borgerhoff Mulder, M. & Richerson, P.J. (2018): Applying cultural evolution to sustainability challenges: An introduction to the special issue, *Sustainability Science* 13: 1–8.

Burman, Å. (2023): *Nonideal Social Ontology*, New York: Oxford University Press.

Claidière, N., Scott-Phillips, T.C. & Sperber, D. (2014): How Darwinian is cultural evolution?, *Philosophical Transactions of the Royal Society B: Biological Sciences* 369: 20130368.

Dahl, R.A. (1957): The concept of power, *Behavioral Science* 2: 201–215.

Daniels, J. (2012): Race and racism in internet studies: A review and critique, *New Media & Society* 15: 695–719.

Epstein, B. (2009): Ontological individualism reconsidered, *Synthese* 166: 187–213.

Epstein, B. (2015): *The Ant Trap: Rebuilding the Foundations of the Social Sciences*, New York: Oxford University Press.

Foucault, M. (1982): The subject and power, *Critical Inquiry* 8: 777–795.

Fracchia, J. & Lewontin, R.C. (1999): Does culture evolve?, *History and Theory* 38: 52–78.

Fracchia, J. & Lewontin, R.C. (2005): The price of metaphor, *History and Theory: Studies in the Philosophy of History* 44: 14–29.

Freelon, D., Bossetta, M., Wells, C., Lukito, J., Xia, Y. & Adams, K. (2022): Black trolls matter: Racial and ideological asymmetries in social media disinformation, *Social Science Computer Review* 40: 560–578.

Frey, W.R., Ward, L.M., Weiss, A. & Cogburn, C.D. (2022): Digital White racial socialization: Social media and the case of whiteness, *Journal of Research on Adolescence* 32: 919–937.

Giddens, A. (1984): *The Constitution of Society: Outline of the Theory of Structuration*, Berkeley: University of California Press.

Glasgow, J. (2009): Racism as disrespect, *Ethics* 120: 64–93.

Haslanger, S. (2000): Gender and race: (What) are they? (What) do we want them to be?, *Noûs* 34: 31–55.

Haslanger, S. (2020): Failures of methodological individualism: The materiality of social systems, *Journal of Social Philosophy* 53: 512–534.

Haslanger, S. (2023): Systemic and structural injustice: Is there a difference?, *Philosophy* 98: 1–27.

Henrich, J. (2001): Cultural transmission and the diffusion of innovations: Adoption dynamics indicate that biased cultural transmission is the predominate force in behavioral change, *American Anthropologist* 103: 992–1013.

Henrich, J. (2004): Cultural group selection, coevolutionary processes and large-scale cooperation, *Journal of Economic Behavior & Organization* 53: 3–35.

Henrich, J. & Broesch, J. (2011): On the nature of cultural transmission networks: Evidence from Fijian villages for adaptive learning biases, *Philosophical Transactions of the Royal Society B: Biological Sciences* 366: 1139–1148.

Henrich, J. & Gil-White, F.J. (2001): The evolution of prestige: Freely conferred deference as a mechanism for enhancing the benefits of cultural transmission, *Evolution and Human Behavior* 22: 165–196.

Jiménez, Á.V. & Mesoudi, A. (2019): Prestige-biased social learning: Current evidence and outstanding questions, *Palgrave Communications* 5: 20.

Kang, J. (2000): Cyber-race, *Harvard Law Review* 113: 1130–1208.

Kroeber, A.L. (1917): The superorganic, *American Anthropologist* 19: 163–213.
Lewens, T. (2009): Population and innovation, in: Krohs, U. & Kroes, P. (Eds): *Functions in Biological and Artificial Worlds*, Cambridge, MA: MIT Press, pp. 243–257.
Lewens, T. (2015): *Cultural Evolution: Conceptual Challenges*, Oxford: Oxford University Press.
Lewontin, R.C. (2005): The wars over evolution, *The New York Review of Books*, 20 October, https://www.nybooks.com/articles/2005/10/20/the-wars-over-evolution/ (Accessed 10 March 2025).
Lewontin, R.C. & Levins, R. (2007): *Biology under the Influence: Dialectical Essays on the Coevolution of Nature and Society*, New York: NYU Press.
Little, D. (2024): Rethinking ontological individualism, *Philosophy of the Social Sciences* 54: 187–202.
Love, A.C. & Wimsatt, W. (Eds) (2019): *Beyond the Meme: Development and Structure in Cultural Evolution* (Vol. 22), Minneapolis: University of Minnesota Press.
Lukes, S. (1974): *Power: A Radical View*, London: Bloomsbury Publishing.
Marwick, A. (2013): Gender, sexuality, and social media, in: Hunsinger, J. & Senft, T. (Eds): *The Social Media Handbook*, New York: Routledge, pp. 59–75.
Mayr, E. (1959): Typological versus population thinking, in: Meggers, B.J. (Ed): *Evolution and Anthropology: A Centennial Appraisal*, Washington, DC: Anthropological Society of Washington, pp. 409–412.
Messias, J., Vikatos, P. & Benevenuto, F. (2017): White, man, and highly followed: Gender and race inequalities in Twitter, in: Sheth, A. (Ed): *WI '17: Proceedings of the International Conference on Web Intelligence*, New York, NY: Association for Computing Machinery, pp. 266–274.
Msimang, P. (2019): Racializing races: The racialized groups of interactive constructionism do not undermine social theories of race, *Ergo* 6: 1–30.
O'Connor, C. (2019): *The Origins of Unfairness: Social Categories and Cultural Evolution*, Oxford & New York: Oxford University Press.
O'Connor, C., Bright, L.K. & Bruner, J.P. (2019): The emergence of intersectional disadvantage, *Social Epistemology* 33: 23–41.
Omi, M. & Winant, H. (1986): *Racial Formation in the United States: From the 1960s to the 1980s*, New York: Routledge & Kegan Paul.
Parker, D. & Song, M. (2006): New ethnicities online: Reflexive racialisation and the internet, *The Sociological Review* 54: 575–594.
Richerson, P., Baldini, R., Bell, A.V., Demps, K., Frost, K., Hillis, V., Mathew, S., Newton, E.K., Naar, N., Newson, L., Ross, C., Smaldino, P.E., Waring, T.M. & Zefferman, M. (2016): Cultural group selection plays an essential role in explaining human cooperation: A sketch of the evidence, *Behavioral and Brain Sciences* 39: e30.
Richerson, P.J. & Boyd, R. (2005): *Not by Genes Alone: How Culture Transformed Human Evolution*, Chicago: University of Chicago Press.
Richerson, P. & Henrich, J. (2012): Tribal social instincts and the cultural evolution of institutions to solve collective action problems, *Cliodynamics: The Journal of Quantitative History and Cultural Evolution* 3: 38–80.
Safarzyńska, K. & van den Bergh, J.C. (2010): Evolving power and environmental policy: Explaining institutional change with group selection, *Ecological Economics* 69: 743–752.
Sawyer, R.K. (2005): *Social Emergence: Societies as Complex Systems*, Cambridge: Cambridge University Press.

Sperber, D. & Claidière, N. (2008): Defining and explaining culture (comments on Richerson and Boyd, *not by genes alone*), *Biology & Philosophy* 23: 283–292.
Valković, M. (2023): Ontological and methodological limitations of certain cultural evolution approaches, *Philosophy of the Social Sciences* 54: 279–301.
Van Doorn, N. (2011): Digital spaces, material traces: How matter comes to matter in online performances of gender, sexuality and embodiment, *Media, Culture & Society* 33: 531–547.
Wimsatt, W.C. (2013): Articulating babel: An approach to cultural evolution, *Studies in History and Philosophy of Science Part C: Studies in History and Philosophy of Biological and Biomedical Sciences* 44: 563–571.
Wimsatt, W.C. & Griesemer, J.R. (2007): Reproducing entrenchments to scaffold culture: The central role of development in cultural evolution, in: Sansom, R. & Brandon, R.N. (Eds): *Integrating Evolution and Development: From Theory to Practice*, Cambridge, MA: MIT Press, pp. 227–323.

6 Rethinking the Concept of "Power" Within Cultural Evolution

Cultural Niches as Relational Social Structures

Karim Baraghith and Finn Jordan

6.1 Introduction[1]

Can cultural evolutionary theory (CET) explain power? Surprisingly, this question has seldom been raised, except for some rare cases of critical authors who share a substantial amount of skepticism about the matter. For example, Fracchia and Lewontin claimed that the evolutionary success of a cultural variant (i.e., a piece of cultural information) depends "crucially on the amount of power behind it" (2005: 22). Likewise, Durham (1991: 182) claims that CET "ignores *structured asymmetries* or power relations, and *coercion*". In this chapter, it is our aim to describe precisely these aspects of power: (a) structured asymmetries and (b) coercion, against the background of CET's framework.

Can power be adequately explained within the framework of CET, particularly through the concept of "prestige bias"? Since the seminal contributions by Boyd & Richerson (1985), it has been recognized that cultural variants associated with prestige tend to proliferate more widely. However, this does not address the full spectrum of power. As demonstrated by Cofnas (2018), while influence and manipulation can indeed be modeled through prestige bias in CET frameworks, this approach – though capturing certain dimensions of power – fails to encapsulate a critical aspect of it, which is fundamentally disconnected from the underlying logic of prestige. The type of power posing a challenge to CET is that which is underpinned by coercion, force, or authority, effectively overriding the typical learning biases in cultural evolution and significantly reducing individual agency in choosing between cultural variants.

Building on the recent discussions surrounding power in CET, this chapter introduces two additional points, augmenting the insights previously offered by Cofnas (2018), Lewens (2015), and Fracchia & Lewontin (2005) in their critical examinations of power within CET.

Firstly, we believe that not just the *absence of individual choice* is a problem (as these authors claim), but also the fact that *coercion and force*

DOI: 10.4324/9781003643845-8

do not seem to follow the logic of evolutionary fitness, according to which a trait increases its fitness by increasing its number, as Henrich & Gil-White (2001) already pointed out. Powerful and coercive individuals are not interested in transmitting their behavioral patterns and being imitated by subordinates; they are interested in obedience and order following. Structurally, a dominant group or a cultural elite might remain powerful and successful even if not too many people are able to join it, that is, their cultural variants are not copied too often. This might be for several reasons: either because the elite actively hinders copying, because there is no interest in copying their behavior, or for other reasons.

Secondly, CET has trouble explaining structural aspects of power. In many cases, power and social asymmetries in human societies cannot be reduced to single individuals free of norms, institutions, or social structures. In this respect, *institutional* or *structural power* needs explanation.

In Section 6.2, we will start by describing the overall richness (and vagueness) of the term "power". We suggest restricting its conceptual richness to two aspects in particular: prestige and coercion. Usually, prestige is contrasted with "dominance" in the literature. However, we chose the term "coercion" instead of "dominance" here, since we want to emphasis on cultural rather than on biological evolution, where dominance hierarchies play a huge role in animal societies. In this way, *we understand coercion as a form of "cultural dominance"*.

In Section 6.3, we will offer possible solutions to the puzzle of dominant agents not being copied. We discuss whether coercion could be seen as a novel selective bias in CE – different from prestige but of the same kind; or whether it is just a biological trait, ultimately, one that does not need cultural copying and imitation because it simply reproduces genetically. We argue for a new interpretation of cultural dominance in terms of r- and K-selection. This dichotomy between K-selected conservatism and r-selected innovation has profound implications for our understanding of cultural evolution. It posits that cultural traits are not merely passed down through generations but are actively selected based on the prevailing social and environmental conditions. These conditions are significantly influenced by power structures within societies, which can either hinder or promote the spread of cultural traits.

Finally, in Section 6.4, we will embed aspects of social structure and explain that the reproduction of structural coercion can be seen as a kind of *cultural niche construction*. We argue that the perpetuation of structural coercion may be viewed as a unique type of cultural niche construction, wherein agents both inhabit and actively create social asymmetries through their interactions. The objective is to establish a quantitative metric of power within this interaction network: the greater the number of other agents (nodes) that a dominant agent can compel to alter their behavior according to its directives, the more power that agent possesses. This imbalance of power in the network's structure, which we understand in terms of a "relational cultural niche", might stay even after the most powerful agents leave and are replaced by others.

6.2 The Challenge: Explaining Power Within CET

6.2.1 *What Is Power?*

"Power" is a vague concept that has a rich history, not only in the social sciences but in political philosophy, as well. Understanding power is of importance for a theory of society that aims for a high explanatory range, such as CET à la Richerson & Boyd (2005), Mesoudi (2011), and Henrich (2016). Power is a result of social hierarchies and helps cultural agents construct and navigate social hierarchies. In most general terms, power can be seen as a "social asymmetry": there are those with power and those lacking it, where power grants those who have it certain abilities and capacities over others. In social settings, power influences decision-making.

While Max Weber (1921/1922) characterizes power as "the probability that one actor within a social relationship will be in a position to carry out his own will despite resistance" (53), according to Hannah Arendt (1958), no single individual can hold power, but rather, power is the potential for acting together (cf. 200–201). According to Dahl's classical definition, "A has power over B to the extent that he can get B to do something that B would not otherwise do" (Dahl, 1957: 202). This definition is quite broad. However, we agree with Cofnas (2018) that it is too broad to serve as a proper working definition within the framework of CET. This is so because every instance of cultural transmission would instantiate a "Dahl-manifestation" of power, which would then result in the semantic emptiness of the power concept. On the other hand, Dahl's canonical definition ("A's potential to get B to do something that B would not otherwise do") is too narrow, since it fails on two more counts. First, it ignores cases in which B would have performed the very same action regardless of A's involvement (e.g., a dedicated tennis player who practices even without her coach's command). Second, by encompassing institutional forms of power (e.g., a tax authority's capacity to enforce compliance), it blurs the analytical focus on discrete, individual agents.

There are lots of fine-grained aspects of power manifested in society. In social settings, we know power as authority, for example, legal power. In democratic societies, for example, state power or authority is typically divided into the legislative, judiciary, and executive branches to prevent the accumulation of too much power. There is also power as dominance of individuals or groups and organizations, as a kind of control over other individuals or groups. Dominance consists of or includes making others do what is in the interest of the dominant agent, rather than representing the interests of other agents. We sometimes distinguish between "soft" power and "hard" power, which differ in whether they do or do not implement force or coercion to be effective. Additionally, classical CET models tend to obscure *indirect forms* of power (such as class or gender hierarchies) precisely because these social facts are epistemically or ontologically more "opaque", see Burman and Pelgrom (Chapter 7 in this volume). Yet such structures often operate without the explicit awareness of those affected, while nonetheless exerting real causal influence.

As you can see, there are many fine-grained forms and aspects of power. It would clearly go beyond the scope of a single article to even list all possible facets of the power concept in philosophy and the social sciences, let alone discuss them against the background of CET. As far as we can see, there is no comprehensive definition of power in the social sciences or political philosophy that enjoys a broad consensus. However, as we hoped to make clear, there are phenomena that bear a certain family resemblance. We do not aim to even try to put them all in a single investigation. Therefore, in what follows, we want to restrict our analysis of power to two sub-concepts or strategies, which we think capture two important functions. These are "*prestige*" and "*dominance*", which diverge in important ways, see Maner & Case (2016). In the following, however, we will speak of "coercion" instead of "dominance", in order to make clear that we only have *cultural evolution* and *"social dominance"* in mind. Why do we choose these two forms of power? We chose them for methodological reasons because we think they can be embedded into the logic of CET. Unlike prestige, which has been widely discussed within CET, coercion/social dominance has not been given sufficient attention.

6.2.2 Power Within CET: Prestige and Coercion

Why should power in general, and coercion in particular, be of any interest to CET; that is, why should it take it into its explanatory repertoire? CET's population-level explanations might, after all, not be "the right tools to use when one wants to understand small-scale systems with networks of heterogeneous actors", cf. Lewens (2015: 5). Since CET mostly looks at large-scale cultural developments and makes quite idealized assumptions, simplifying the behavioral repertoire of cultural agents to a large extent, coercion might be too small scale a phenomenon to be of any explanatory value for a population-based approach like CET.

Still, phenomena of power are omnipresent in all human societies and have the potential to put executive forces of large-scale and long-term cultural developments in the hands of only a few actors. Very often, power also becomes institutionalized (see Lewens, 2015: 136), and it seems that CET has not much to say so far on institutions, even though they are of tremendous importance to cultural evolution, see Currie et al. (2021) or Hindriks (Chapter 4, this volume). Power structures are important, and we think that CET will do well to start dealing with them.

If we define power as a potential of A to make B do what is primarily in A's interest, or what B would not do in the absence of A, then A has power and B does not. In this most abstract reading (which does not even capture all facets of power), power can be described as a particular *causal* influence that A has on B, and we think that this influence can at least appear in two forms. In modern societies, we do encounter both, and it is likely that both may serve *cultural functions*, as cultures are founded on power systems. A lot of information, opinions, and options for action are "*constrained in all sorts of ways by the (often hidden) exercise of force and authority*", cf. Cofnas (2018: 305).

The (evolutionary motivated) distinction between *dominance* and *prestige* was already popularized by Henrich & Gil-White (2001), see Figure 6.1. Additionally, in evolutionary psychology, these two aspects of power represent two different strategies to navigate social hierarchies and to maximize an agent's utility (see Maner, 2017). Prestige and dominance have their own distinct psychology selected for by distinct evolutionary pressures (see Henrich & Gil-White, 2001). It was assumed in early evolutionary psychology that (cultural) prestige evolved from (natural) dominance, that prestige is – so to speak – a cultural descendant or a *cultural*

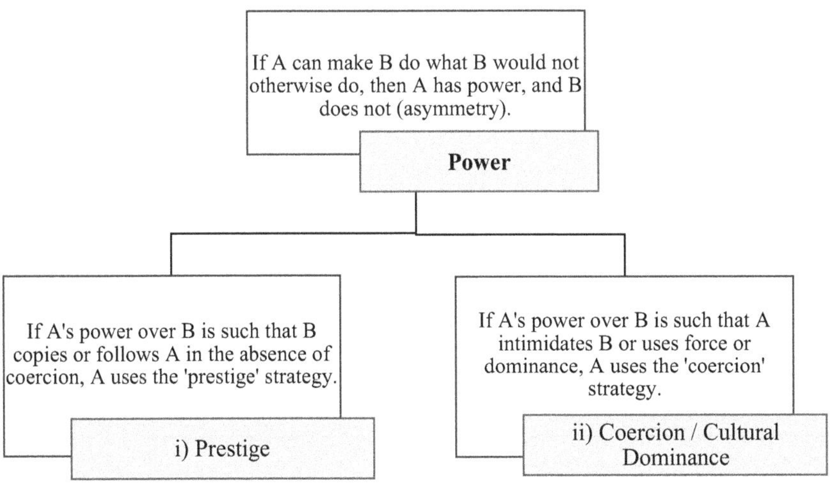

Figure 6.1 Two important forms of power. While present CET has minor issues explaining (i), there are problems explaining (ii).

exaptation of dominance, see Barkow (1989). Our non-human ancestors executed dominance, we execute prestige, but both serve the same function ultimately: an increase in biological fitness – a classical explanatory move in "high church" evolutionary psychology. But why should selection transform dominance into prestige? And why are both forms of power still present in modern societies? It may be true that dominance is a phylogenetically much older form of social asymmetry, and that prestige is much younger, but that does not mean that there is a direct line of descent connecting them from a cultural evolutionary perspective. Therefore, we do not have to agree with Barkow (1989). There are even different personality profiles associated with dominance and prestige (see Maner, 2017). While dominant individuals have a tendency for narcissism or Machiavellianism, prestigious ones are high in agreeableness, self-esteem, and fear of negative evaluation (see Maner & Case, 2016).

In dominance hierarchies, subordinates usually seek to avoid superiors (e.g., out of fear). This is not true for prestige hierarchies, where contact is actively sought by subordinates. They are even willing to pay certain costs in order to be in contact with, learn from, or have the attention of superiors. Henrich & Gil-White (2001) explain this phenomenon of individuals paying costly deference with their "information goods theory", which can make prestige an evolutionary stable strategy. The authors proposed an account of the cultural evolution of prestige that is now well established and classic. They contrast dominance and prestige hierarchies, pointing out specific differences between these two strategies. The following table from Henrich & Gil-White (2001: 171) lists the most important ethological aspects of these two hierarchies.

	Prestige	*Dominance*
Low-status individual		
1) Approaches often	Yes	No
2) Prolonged stares	Yes	No
3) Occasional challenge	No	Yes
4) Fears high-status individual	No	Yes
High-status individual		
5) Charges and attacks	No	Yes
6) **Is frequently imitated**	**Yes**	**No**
7) Has more biological offspring than average group member	Yes	Yes
8) Received gifts/services	Yes	Yes

According to point (7), it becomes clear that both forms of power grant superiors more biological offspring. But this is a biological argument. What about the cultural advantages of prestige and dominance? The most important point for our purposes is point (6): *behavioral patterns executed by prestigious individuals get copied, but not so in the case of dominant individuals.*

Note that in modern human societies, prestige and dominance are often blended in complicated ways. Dominance is often interwoven with prestige. The more prestigious a superior becomes, the more likely people are willing to tolerate or even expect dominant behavior from her – at least in certain legitimate situations. For the time being, however, let us leave this complicated interplay between the two strategies out and focus on each pattern individually to gain a better understanding.

In the setting of CET, the reproductive element of prestige – given as "prestige bias", that is, what makes prestigious individuals more likely to be copied by others – is well known and studied, even though it remains subject to lively philosophical debate (see Chellappoo, 2020). Boyd & Richerson (1985) described prestige bias (a model-based bias) as a specific force acting on cultural selection, one that makes a cultural agent copy or follow renowned and high-status individuals, often those with high success. For example, new innovations are more likely to become successful on the market if they are produced or advertised by already successful individuals. A deeper analysis of prestige bias in CET is given by Henrich & Gil-White (2001), Henrich (2016), and Chapter 8. According to their evolutionary story, prestige bias is adaptive since prestige indicators (such as deference) act as cues for competence. Gaining competence (or just getting copied frequently for whatever other reasons) then leads to a higher chance of becoming prestigious yourself. Prestige is, no doubt, a form of power.

As Mesoudi (2011: 75) points out, even linguistic changes in dialects are often driven by high-status individuals in a community, as experiments in sociolinguistics indicate. This is prestige bias: the more prestigious you are already, the more likely you will be copied, or your innovations will spread. People tend to follow prestige, just as they tend to follow success. However, success and prestige are not the same, even though they often come hand in hand. Prestige can be an indicator of success, but that is not necessarily so. Sometimes, cultural agents do not have direct *access* to the success rates of other individuals and simply rely on their reputation or prestige. Therefore, prestige is more indirect than success but easier to infer. If no deception or mistakes are involved in the cultural process, and if the cultural community indicates and values certain successful traits *as* prestigious, prestige can be a reliable indicator of success. Note that prestige bias, as a specific force acting on cultural selection, has the tendency to result in a kind of "runaway" process in certain domains of society. Consequently, Boyd & Richerson (1985: 259–271) present a model for the runaway selection of prestige. They draw an analogy to sexual selection in biological evolution, namely the famous "paradox" of the peacock tail's selection, which gave Charles Darwin a hard time thinking about and was finally solved by Ronald A. Fisher. In this reading, cultural prestige is equivalent to *sexual preference* in biology. Just like in the case of sexual selection, markers of prestige can exaggerate in value, such as "house size", Mesoudi (2011: 76). That is, the selective tendency of prestige bias for larger and larger houses as an indicator of power can lead to exceptionally large and expensive mansions for high-status cultural agents, equivalent to exceptionally showy peacock tails. Other examples are possible, such as "small family size" (Mesoudi, 2011: 81). Whereas natural selection generally favors large family sizes, prestige-biased cultural selection may favor small families because their members have more time to devote to and invest in their careers, free time, and social status.

All cases of prestige bias described so far are relatively clear and well-reasoned by CET. *They do not explain, however, how dominance and force work*. Cultural "power", in the sense we defined it, can by no means be reduced to prestige. Coercion, intimidation, and force are alive and well in all human cultures and societies – from hunter-gatherer societies to globalized internet telecommunication networks – and are by far not restricted to primate societies. Coercion describes the ability of a person, a group, or an institution to make others do what is in their (and not necessarily the other's) interest. In this sense, powerful dominant individuals, unlike prestigious ones, will not be idols or models to be imitated or willingly followed. To the contrary, a dominant individual or group makes others do what is in their best interest, often by executing a specific behavior that leads to obedience or order-following.

CET has a hard time explaining force or dominance. There is surprisingly little literature on force/coercion and CET. Even if some authors write about it, they do not embed coercion. A prominent example is O'Connor (2019), who addresses "power", but only in the sense of the emergence of inequality (unfairness) in sharing a resource. However, as Cofnas (2018: 300–301) wrote:

> force is the form of power that poses the most obvious (apparent) challenge to the models. If cultural variants are imposed on people because the option to reject them is taken away by force, then it seems that standard cultural evolutionary models cannot explain what happened. . . . Learning biases might explain why people choose to accept the authority of a particular leader in the first place, but they would not determine which variants subsequently spread.

Both prestige and dominance will create deference, but only prestige creates copying. This does not mean that powerful individuals cannot force other cultural agents *to copy* them. A superior might ask or order one or several subordinates to exactly imitate her behavior. *Powerful individuals can coerce conformity*. However, the reason why they will do so lies not in the superior's prestige, but in their ability to use coercion successfully. Again, coercion is in need of explanation, and if CET's only machinery to deal with such a case is prestige, it is incomplete.

6.2.3 *The Logic Behind Coercion: "Don't Do What I Do, Do as I Say!"*

Coercion has not really been incorporated into the explanatory repertoire of CET, and probably for good reasons. A cultural fitness-based explanation is problematic if there is seemingly no reproduction involved. Why is this a problem at all? It is because the theoretical tools developed by CET seemingly do not have anything to say about this kind of non-reproductive form of power, its distribution in a society, and the large influence that it certainly has on almost all cultural processes.

Fracchia & Lewontin (2005), Lewens (2015), and Cofnas (2018) are right to notice that CET is not able to deal with this kind of power, but we think that this is the case for different reasons. We think that one reason why CET in its present form does not model force and authority is not necessarily a commitment to methodological individualism (as these authors claim); it is because dominance and force seem to reverse the logic behind CET explanations (or any kind of evolutionary fitness-based explanation). Dominant individuals (or groups) can bring about "a lasting change in the behavioral traits" within a cultural population. This latter formulation is taken from Ramsey (2013), who provided an insightful definition of culture, one that considers a variant-based as well as an organism-based view on cultural fitness.

$A, B, C, \ldots$ = Cultural agents
p = Prestige
c = Coercion
v_i = Any cultural variant/behavior
$A(p)_{\iota}$ = Agent A uses the prestige strategy
$A(d)_{v_i}$ = Agent A uses the coercion strategy

Figure 6.2 Schematic illustration of the structural differences between prestige and coercion. While a cultural variant v_i is copied in case (1) because of prestige bias (p), there is no copying at all in case (2), only the causal influence of a dominant individual $A(d)$.

Let us take a more detailed look at power within CET. Figure 6.2 visualizes the differences between cultural interactions that are subject (1) to prestige and (2) to coercion. Arrows indicate social interactions between agents, that is, agents using specific social strategies toward other agents. Note that in this chapter, we understand prestige and dominance in a twofold way: first, these different forms of power describe a certain *potential* of agents; and second, they describe strategies to navigate social hierarchies. Understood in a dispositional sense, power is a *potential* of an agent to enforce her cultural variants with either coercion or prestige. If she *actually* executes these enforcing variants, she uses the coercion or prestige strategy.

A cultural trait v_i that is neither enforced by coercion nor prestige will be less frequent in the next generation than it would be under prestige enforcement. v_i enforced by p (written as: p_{v_i}) will spread more often. This is not the case for coercion, where an individual *simply enforces a certain behavior in other individuals*, even in the absence of copying the behavior or the strategy. In case (2), an individual A is executing v_1, which is backed up by d. A influences three other individuals (B, C, and D) via a direct order or a threat. B, C, and D each then perform different actions (v_2, v_3, v_4). There is no need for these actions to be replicas of v_1 or even resemble it in any noticeable way.

Think of a military commander giving three different orders to three different subordinates ("You build the camp!", "You take patrol!", and "You scout the area!"), each of them then executing the order by performing very different behaviors. This may be known as the "*Don't do what I do, but do as I say*" logic. Even if there is no observable resemblance between v_1 and the other actions, we would say that they are crucially influenced by A, and if they increase in fitness (frequency), this is also because of A. To make the point even stronger: in the case of a very powerful A, (v_2, v_3, v_4) only *exists* because of A, that is, A is their cause.

Two questions become pressing. First, if coercion is not copied culturally, how come it is so present in society? Second, if CET cannot really offer a copy-based explanation of a case like (2), what is to be made of that?

6.3 Possible Solutions

In what follows, we try to navigate a space of possible solutions to the problem of cultural dominance/coercion. Lewens (2015: 138) offers three options for CET to deal with the "Durham, Fracchia, or Lewontin critique" (DLF) that CET cannot really deal with "power". He calls them (1) restriction, (2) presumption, and (3) reduction. Accordingly, Cofnas (2018) offers three options on how to deal with that (cf. Cofnas, 2018: 304).

One could (1) restrict the use of CET models to times and places where they are not driven by DLF power relations. This approach undermines any modeling applied to modern social science. Lewens suggests an application of CET models to early hominin history. Based on Boehm (1999), Cofnas (2018) objects to this line of thinking on the basis that hunter-gatherer societies were not free of dominance relations, and that even egalitarianism was collectively enforced as "reversed dominance hierarchy". Alternatively, you could (2) incorporate power as a fixed background condition against which individuals interact and use CET to track change resulting from those interactions. This somehow ignores the DFL challenge and a primary factor for cultural evolution. Or you can (3) try to reduce systems of power to small-scale interactions dealt with by standard CET, the favored approach of evolutionary modelers.

Note that Chellappoo (Chapter 5, this volume) argues that *neither of these replies* to DLF adequately restores CET's ability to account for power: (1) which confines CET's application to contexts without significant power imbalances, would exclude precisely those domains, such as digital media, institutional governance, and environmental policy, where power inequalities are most consequential, thus unduly narrowing the theory's scope; (2) which treats institutions and social structures as fixed background conditions and neglects the dynamic, recursive interplay in which individuals both shape and are shaped by those very structures, leaving CET blind to the ongoing reproduction and transformation of power relations; finally, (3) which seeks to explain structural hierarchies purely as the aggregate of individual interactions, fails because it cannot capture emergent, system-level phenomena (e.g., institutional racism) that do not supervene on individual properties.

Our own approach (see Section 6.3.3 below) goes in the direction of (3) to answer DLF, but we intend to expand our CET explanation of coercion with a network-based social structure. In other words, we propose a *fourth option* next to restriction, presumption, and reduction, that is, (4) extension.

6.3.1 Option 1: Coercion as Selective Bias in Cultural Transmission

It seems that Cofnas (2018) chooses option (1): restriction, because he restricts CET models to certain non-moral domains of social interaction. It might be that dominance can be reduced to a social learning bias in CET. At first sight, it appears to be a force of cultural selection. Since coercion, just like prestige, can be understood as a characteristic of certain individuals, we may be inclined to think of it as a model-based bias. However, as we have illustrated, cultural variants exhibited by coercive individuals are not copied more frequently. Rather, it implies the enforcement of cultural variants that are not necessarily exhibited by the dominant individuals themselves and are often not further reproduced as soon as the respective threat has been lifted. But is it possible to include coercive cultural reproduction in terms of a content bias? As Cofnas (2018: 305) suggests, we might regard the negative, coercion-induced incentives presented for adopting a cultural variant as a feature of the environment. The threat of force itself, however, is not part of the respective variant's content – it is rather a contextual factor of a cultural variant in a specific environment, making a potential bias appear much closer to the category of context biases. Cofnas circumvents this problem by pointing out that preferring cultural variants whose adoption is not accompanied by significant costs might constitute a general "content bias". We could then say that individuals affected by coercion merely choose which cultural variant to acquire based on the provided incentives. Consequently, according to this understanding, if the dominant individuals do not exhibit the enforced cultural variant themselves, acquiring it may be conceived as a cost-sensitive, guided variation based on environmental aspects, namely the presented threats of force.

Yet, as Cofnas mentions, this framework may be unsatisfactory since we leave coercion unexplained, making "what may often be the leading cause of transformation" (see Durham, 1991: 182), a background condition in need of external inputs regarding which individuals are affected, what cultural variants are enforced, and how high the costs for resisting would be.

6.3.2 *Option 2: Coercion as an Entirely Biological Trait*

As we mentioned earlier, dominant individuals have a higher chance of having more biological offspring. A sociobiological explanation of the fact that coercion (cultural dominance) is present in culture might be that it is simply genetically inherited. It could be a case of "evoked culture", see Mameli (2008). If this were the case, coercive behavior would be part of our human genetic package, and all cultural instantiations of it could be seen as *exaptations*. Many proponents of CET seem to go in this direction (at least implicitly), highlighting that dominance can be found already in so many non-human animal societies, and that is phylogenetically much older than prestige (see, e.g., Henrich, 2016, Chapter 8, or Henrich & Gil-White, 2001).

On the face of it, we do not think that this is a very plausible way to go, and it does not satisfactorily explain cultural dominance and power. We do not deny that dominance is present in our non-human cousins and that it may have a strong biological component in humans, as well. Dominance has been connected to hormonal features, especially a high level of testosterone; see Mazur & Booth (1998). Mileva et al. (2014) even linked dominance physiologically to a wider facial width-to-height ratio, a morphological cue associated with greater testosterone, aggressiveness, and unethical behavior. Coercive (or aggressive) behavior of a cultural agent can stand in a causal connection to hormonal reactions of its biological organism, and it might even be that specific hormonal distributions have different genetic components; that is, there is a biologically inherited element in dominance behavior. For a more extensive overview of the biological features of dominance, see especially Maner (2017). However, coercion, as we defined it above, is a deeply cultural phenomenon, and cultural dominance is simply not reducible to the biological dominance that we find, for example, in our non-human ancestors. If it were so, we would expect every person who ever executed cultural dominance, that is, made other cultural agents follow and obey her orders, to have a biological heritage of dominance. Most certainly, this is not the case, especially when dominance becomes structural, that is, becomes a feature of social networks.

6.3.3 Option 3: Coercion and the Logic of Cultural Fitness: Growth, Reproduction, r and K-Selection

We think there is another possibility to incorporate power into CET's framework.

Assume an elite (N) in a society that is very powerful; that is, cultural information passed on among them has a high probability of bringing about "a lasting change in behavioral traits" (Ramsey, 2013). This lasting change also affects many other members of society. How can N, in fact, be successful even if the number of N's cultural variants (as well as their relative number in society) is very low? N's information or behavior could last for decades or even centuries, even if they never reached high frequency or their cultural traits were stored in many heads. In other words, N backs up their variants with power, but this power (cultural dominance) will not proliferate. How can we embed such a logic into cultural evolutionary thinking? Is that even possible?

Let us have a closer look at biological evolution and see whether we find some analogous phenomenon to this cultural dominance. A single biological organism can be quite successful (you might even say "powerful", but that might overstretch the meaning of this concept) during its lifetime. It can have a large territory, acquire lots of sustenance, grow very large in size, etc. In other words, it can be very dominant in a certain environment – equally toward conspecifics as toward its whole ecosystem. It can, for example, stand on top of the food chain, or it can have privileged access to water, nutrition, etc. Think of an elephant bull approaching an almost dried waterhole: all other animals (including top predators such as lions) will grant him privileged access out of fear because he is so "powerful". However, this dominance is completely meaningless for the elephant's genetic fitness unless the elephant reproduces more often. As in so many instantiations of analogical reasoning between biological and cultural evolution, we must carefully distinguish "growth" from "reproduction", as Ramsey & De Block (2017) show for the case of cultural fitness. An elephant has a lot of somatic cells due to its size, but that does not mean that it has a lot of offspring. Large animals (like whales or elephants) have very slow reproduction rates, and when they do reproduce, they usually only have one young. This is simply due to environmental and physical conditions, mostly due to limited resources. A female elephant *would* have 15 young at the same time if she *could*, that is, if her physical conditions and the environment allowed it. But the cost of getting as large and environmentally dominant as an elephant is a very slow reproduction rate. A mouse can have 15 babies at the same time and be pregnant again some weeks later.

Evolutionary ecology captures this with two concepts, denoted *r* and *K selection,* see Pianka (1970). *r* comes from the term "rate" and *K* from the German "Kapazitätsgrenze" (capacity limit). The first selective strategy *r* is to have as many offspring as possible, the second *K* is to only have a few offspring but ensure their survival through long maternal care, large size, etc. In evolutionary terms, an elephant follows a "K strategy" and a mouse follows an "r strategy".[2] It seems that by recognizing this, we have found our biological analogue for power: *power given as cultural dominance follows a "K-strategy" while power as prestige follows an "r-strategy".*

That means that the behavior of prestigious individuals (strengthened by prestige bias) will be reproduced frequently by many other agents, while this is rarely the case for dominant individuals' behavior – such cultural variants might stay low in number (due to a limited capacity of the environment to sustain many powerful individuals at the same time) but still yield a tremendous effect on cultural evolution. Note that these thoughts are pure speculation, so far, but let us see where this speculation leads us.

To do so, we will take a look at the mathematical structure of r and K selection. In the case of an evolving cultural population (with discrete generations), we have:

$$N(t+1) = N(t) + \mathrm{rN}(\mathrm{t}) \tag{1}$$

where t describes a point in time (or a generation), and r is the individual reproductive rate of specific variants out of a set of possible variants, $V = \{v_1, v_2, v_3, \ldots v_n\}$, within a population N. We can also describe the population's change (now in a continuous manner, which is a bit easier for our purposes) as:

$$\frac{\Delta N}{\Delta t} = \mathrm{rN} \times \left(1 - \frac{N}{K}\right) \tag{2}$$

N is the population, $\frac{dN}{dt}$ describes the population's growth relative to time, r is the maximum growth rate, and K is the carrying capacity of the local environment. Competitive ability can be increased by aiming for a K strategy. If a population has already reached the capacity limit (K) of the habitat, frequent reproduction is hardly worthwhile. Rather, the number of individuals will now remain almost *constant* over a long period of time. Following a K strategy (coercion) would then become a special cultural case, where capacity limits are reached for one reason or another.

Generally, the higher the amount of dominance within one individual or a small group of individuals, the lower the reproduction rates of the power-enforced variants v_i. On the other hand, the more cultural evolution builds on this strategy, the more likely these *powerful* cultural agents will survive. In the case of an increase, r approaches the maximum number of individual variants (more variants) and K approaches the maximum carrying capacity of the environment (more dominance). Therefore, if K gets very high, the second term $\left(1-\frac{N}{K}\right)$ approaches 1.

To capture this idea within CET, let us introduce a new parameter d, which serves as a weighting for the second term of the equation. Let d_{v_i} be a variant under the enforcement of dominance, then:

$$\frac{\Delta N}{\Delta t}=\left(rN\times \boldsymbol{d}_{v_i}\right)\times\left(1-\frac{N}{K}\right) \tag{3}$$

The first product – now in brackets – can become large in two different ways: either if r grows (the number of variants of one sort), or if d grows (the amount of dominance behind one variant). Recapitulating Figure 6.2, d indicates the amount of dominance of a cultural agent.

Attached to a certain variant v_i, d indicates that this variant may not become large in number but has a high impact on *other* variants' proliferation. Executing force or coercion means making other people do what is in your interest. In a way, the parameter d does also measures how "resistant" an agent is toward being forced by other agents. If the numerical value d of agent 1 is higher than that of agent 2, then agent 1 is resistant to agent 2's manipulation.

Let us make an example: a superior agent might give a specific order (v_1) to 20 subordinates ("Soldiers, present arms!"), who will all react to that order by obeying and executing v_2. In this case, v_1 is backed up by d, while v_2 is not. Even though v_2 has now a temporal "fitness" of 20 and v_1 only that of 1, v_1 has a much more lasting effect. The higher d_{v_i} gets, the smaller r can be. That means, the more your variants are backed up by dominance, the fewer of them you will need to have a lasting effect on other variants. A standard population dynamical approach that only looks at the relative frequency of a variant compared to other variants will not capture this phenomenon.

The analogy between prestige and coercion on the one hand, and r- and K-selection on the other, leaves room for some speculations that would be most interesting to test empirically: cultural individuals who adopt K-selected strategies might tend to prioritize the maintenance of existing social hierarchies and cultural stabilities. Their preference for dominance manifests in efforts to perpetuate traditional cultural practices and social institutions, often at the expense of innovation and experimentation. This conservative approach suggests a risk-averse attitude aimed at preserving their standing and ensuring ongoing stability within their social spheres. Conversely, individuals with r-selected strategies, characterized by high reproductive rates and rapid maturation of their behavioral variants, inject a dynamic element into cultural evolution. These individuals, often holding positions of high prestige, are the progenitors of cultural change. Their willingness to engage with novel ideas and practices, coupled with a propensity to take risks, might position them as catalysts for cultural innovation and diversification.

This dichotomy between K-selected conservatism and r-selected innovation has profound implications for our understanding of cultural evolution. It posits that cultural traits are not merely passed down through generations but are actively selected based on the prevailing social and environmental conditions. These conditions are significantly influenced by power structures within societies, which can either hinder or promote the spread of cultural traits.

6.4 Cultural Niche Construction: The Emergence of Structural Power

> Varys: "Power is a curious thing, my lord. . . . It's a riddle without an answer, or rather, too many answers. . . . Some say knowledge is power. Some tell us that all power comes from the gods. Others say it derives from law". . . . Tyrion: "Did you mean to answer your damned riddle?" . . . Varys: "Here, then: Power resides where men believe it resides. No more and no less".
>
> – George R.R. Martin, A Clash of Kings

In this last section, we will inquire into the reproduction of cultural dominance or coercion in more detail. From time to time, dominance must also reproduce culturally, even though it does not follow the r-strategy (otherwise it would not be present in all known human societies). This may be due to different social factors, and also because (dominant as well as other) agents simply have a limited lifespan. When they pass away or retire, their power is usually transmitted to other agents, often within the elite itself.

Formally speaking, if cultural evolution builds on *d* to increase the overall value of the right side of Equation (3), that does not mean that *r* must be zero. But unlike prestige, dominance does not proliferate by the number of its replicants alone. We think that cultural dominance is mostly reproduced within a *social structure*, and the mechanism of how such a structure is constituted can be subsumed under *cultural niche construction*.

6.4.1 *Power as a Social Structure*

There is a crucial aspect of power that we have not addressed so far. In many cases, power and social asymmetries can be *structural*. Instead of paraphrasing this matter of fact with the help of different social theories, let us explain what we mean by social structure with the help of Geertz's (1973) example of sheep stealing. Geertz argues that individual beliefs, prospects, and preferences (key ingredients for a methodological individualist's explanation) are insufficient to comprehend the phenomenon of "stealing". This is so because stealing can only be stealing in a specific social context or *social structure*, for example, the existence of "private property". Removing a sheep from this "property" (on which there must be a certain amount of agreement in a given culture) is a crime only in light of the existence of this structure; without such an agreement/institution, it would be an action like any other.

This is equally true for power and power structures, but what is the ontological status of such a social structure? In the social sciences or sociology, we find, for example, the distinction between (a) "institutional structure" and (b) "relational structure"; see López & Scott (2000). (a) is seen as a set of normative patterns which guide cultural agents' expectations toward each other, ensuring the endurance of social relations, while (b) comprises the social relations themselves, understood as patterns of social interconnections or interactions between the agents. Within the social science literature, these two different readings of social structures have often stood in mutual tension and conflict. However, for our purpose, we suggest understanding the social environment – the cultural niches – as relational structures, that is, the interactions of the agents themselves. In this sense, social structures are networks of social relations, see also Haslanger (2023).

It is important here that Fracchia & Lewontin (2005), Lewens (2015), who relies on them as well as on Lukes (1974), and also Cofnas (2018) agree that CET's leading paradigm – "methodological individualism" – poses a (if not *the*) severe problem in fully explaining power. Even methodological individualists might doubt that CET is able to explain power, but methodological individualism alone certainly fails to explain it, and it is argued that CET's focus on methodological individualism prevents the theory from taking "networks of power" and social structure into account.[3] According to CET, cultural transmission is executed by individual agents, and these individual agents have individual preferences, which can be influenced by biases, for example, by prestige. As Fracchia & Lewontin (2005: 23) write:

> Because of its inescapable methodological individualism and the methodological randomness it spawns, the selectionist paradigm produces a massive blind spot toward the logic of social relations and socially structured antagonisms; it reduces the social conflict produced by those antagonisms to the indifferent difference of competitive selection.

We do agree with this observation, but individualism, as Fracchia & Lewontin (2005) attribute it to CET, might not be "inescapable" for CET, as the next section will show.

6.4.2 *Reproduction of Coercion as Relational Niche Construction Within Networks of Interactions*

Can we make sense of these structural or institutional aspects of power within CET? One way to see it is that both dominant and submissive individuals *create* a social context or environment that facilitates power execution. The idea is that powerful agents (together with their submissives who are also active agents) do *cultural niche construction* in the sense of Laland & O'Brien (2011). In our case, the cultural niche of an agent consists of other agents and their social interactions, and niche construction means altering these interactions.

In defining the niche in such a way, we do rely on a tripartite conceptual distinction by Aaby & Ramsey (2022), who recently distinguished (i) constitutive, (ii) relational, and (iii) external niche construction. (i) describes an active engagement of an organism with environmental factors, (ii) the altering of an organism's behavioral relations to other organisms, and (iii) ecological engineering, i.e., altering the physical environment itself. As the following table from Aaby & Ramsey (2022) shows, all three kinds of niche construction can be studied concerning their intra- and intergenerational effects, that is, from a developmental and evolutionary perspective.

	(i) Constitutive NC	*(ii) Relational NC*	*(iii) External NC*
Intragenerational	Phenotypic plasticity, behavioral innovation	Behavioral plasticity, relocation, social learning	Ecological engineering (web or nest building), external alterations
Intergenerational	Parental effects, including epigenetic inheritance	Behavioral/social and cultural inheritance	Ecological inheritance

Three kinds of niche construction (NC) and their intra- and intergenerational mechanisms, after Aaby & Ramsey (2022).

In this chapter, and in the case of power structures that we want to describe here, we understand NC as (ii) relational since we aim to explain the reproduction of dominance in terms of cultural interactions in a network of agents. The respective niche is a specific cultural environment, a socio-structural background, which supports and incorporates social asymmetries. In our context, we may also speak of *structural coercion* or even *authority*. The emergence of "relational structure" (see López & Scott, 2000) is analogous to "relational niche construction" (see Aaby & Ramsey, 2022).

Such a structure can be represented as a social network. In actual societies and cultural institutions, the social structure is complex, consisting of hundreds or thousands of interacting individuals. For reasons of simplicity, however, we only assume a network of five agents. This is depicted as a finite directed diagraph G, that is, a graph with bi-directed edges between the nodes. Formally, any graph is determined by the set of its nodes/elements V and the edges/connections E between nodes. We write: $G=(V,E)$. Let us assume a social structure to be a set of agents $V=(1,2,\ldots,n)$ and their social relations. The agents are represented by the nodes of a graph, whereas the interactions between them are represented by the edges $E=(v_1,v_2,\ldots,v_n)$. The interactions are the cultural variants. All interactions are mutual and symmetric, that is, we are neutral concerning their diachronic or synchronic order. Since the situation is entirely symmetric, we do not need to implement a time difference (as, e.g., in a natural dialogue) between the mutual interactions. We will also not inquire here about how this toy network emerged in the first place and will take its specific structure as given.

Let us interpret the interactions as cultural variants. In Figure 6.3, any edge $v_i \in E$ is a directed (i.e., guided) interaction between two agents. For example, agent 4 interacts with agent 5 via v_5, while agent 5 reacts via v_6. So far, all interactions are symmetric and of equal value; that is, we assume a completely *unweighted graph*.

This is a quite unrealistic and idealized state of affairs, since we assume that power asymmetries are ubiquitous in social situations. Therefore, let us now introduce weightings and change our graph into a *directed, weighted graph*. Some variants are enforced by dominance d, others are not. Agent 4 enforces all her actions by d, where $(0 \leq d \leq 1)$, that is, we assume d to be a conditional *probability value*, conditionalized on agent 4. Formally speaking, she weights all directed edges to adjacent nodes, that is, nodes in her direct neighborhood, which are (v_4,v_5,v_{10}) in our example; see Figure 6.4.

Therefore, she forces all other agents who interact with her to *change* their variants toward compliance with 4, as we described in Section 6.3.3.

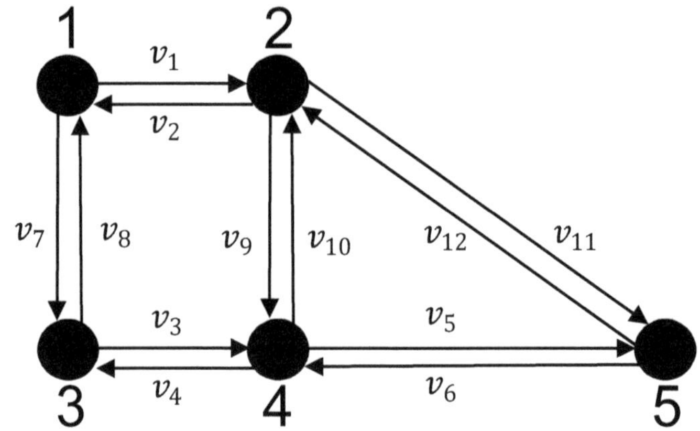

Figure 6.3 A random, unweighted social network. All interactions are mutual, and there are no social asymmetries.

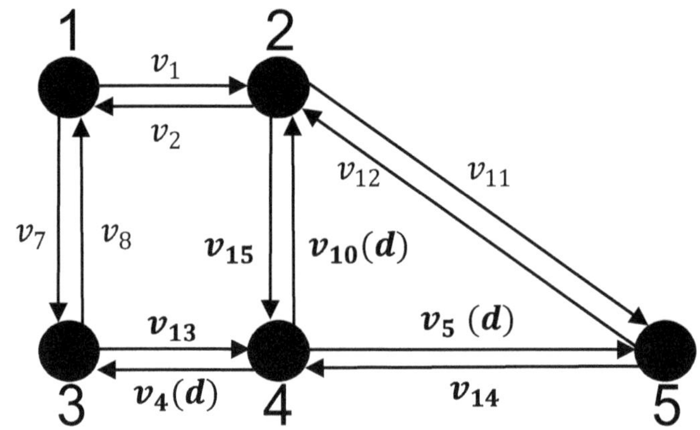

Figure 6.4 A particular agent (4) in the network backs up her relational variants with a dominance weighting, thereby forcing all adjacent agents to alter their behavior.

With probability d an agent changes another agent's variant into a different variant, we write: $d: v_n \to v_m$. In our case, we have $(d|4): v_3 \to v_{13}$, $(d|4): v_9 \to v_{15}$, and $(d|4): v_6 \to v_{14}$. It means that, with probability d, agent 4 changes v_3 to v_{13}, $v_9 \to v_{15}$, and so on. Note that this change takes place in agent 3, who obeys agent 4. After that, the actions of agents 2, 3, and 5 are restructured in favor of 4. A social asymmetry emerged.

Agent 4 actively changes her cultural niche (consisting of agents 2, 3, and 5) in her favor, such that 4's variants will survive but not so the former variants of agents 2, 3, and 5, that is, (v_3, v_9, v_6), which have been repressed from the evolutionary game by agent 4 and went extinct.

We can embed $d : v_n \to v_m$ into equation (3) of the last section. Let us define:

$$d(\#v_n \to \#v_m) := \#v_d$$

that is, the number (#) of variants that change because of d. We then arrive at:

$$\frac{\Delta N}{\Delta t} = \left(rN \times \frac{\#v_d}{d_{v_i}} \right) \times \left(1 - \frac{N}{K} \right) \qquad (4)$$

The probability value d_{v_i} is now relative to the change in the actual number of variants that change into other variants. The more other variants change (i.e., the more agents a dominant agent controls), the higher the term $\frac{\#v_d}{d_{v_i}}$ gets.

Since our objective in this chapter is to establish a quantitative metric of power within this interaction network, we define the *strength of power* by $\frac{\#v_d}{d_{v_i}}$. The greater the number of other agents (nodes) that a dominant agent can compel to alter their behavior according to its directives, the more power that agent possesses.

It is important to note that this niche (the social power structure), which consists of a dominant agent and other agents' interactions, is a culturally distributed phenomenon. To explain this phenomenon, methodological individualism reaches its limits because structural dominance/coercion is a *cluster* of interactions. It is not reducible to the dominant agent or his cultural variants alone, and neither is structural dominance itself. The other agents in the network must comply and act in accordance.

This becomes even more vivid if we assume the following next step; see Figure 6.5. Suppose agent 4 (the dominant one) leaves the network for one reason or another. She might pass away, retire, or fall victim to a political coup. If we take her out of the network, a power vacuum emerges.

What happens to the other agents once agent 4 is gone? We see two options here: either the network collapses and completely reconfigures, or (and this is the more interesting case) the structure persists, and the powerful individual is just being replaced by another one.

In the latter case, those agents who complied with agent 4 and even changed their interactions in favor of her will stick to that new behavior. Agents 2, 3, and 5 retain the cultural variants imposed on them previously by agent 4. They have *dispositions* toward acting in this direction because

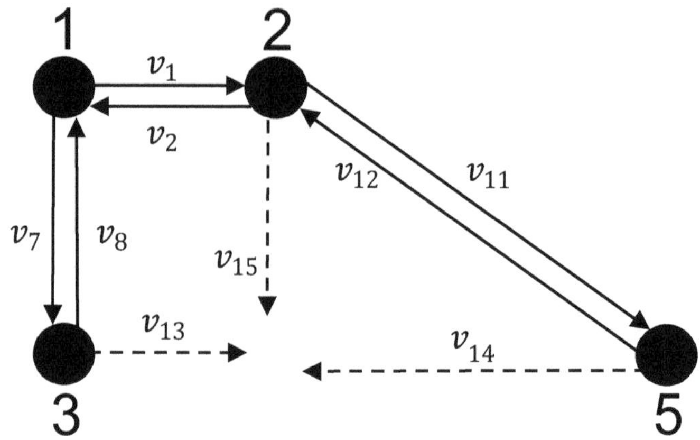

Figure 6.5 The dominant agent has left the network, but the other agents still hold the disposition to act as if she were still present. A power vacuum emerged.

they did so in the past, and they keep these dispositions even in the actual absence of agent 4, even if these dispositions do not manifest directly. So, the niche (the slot/power disposition/*social role*) remains because "*Power resides where men believe it resides*" – and they believe it lies at precisely this position in the network.[4]

Suppose that in the next step, another agent 6 will fill the power vacuum and the "institutionalized position", as shown in Figure 6.6. She might have gotten there by succession, by right, by a general, free, and equal vote of the people, by chance, or because she killed all her potential rivals. That does not matter for our analysis. What matters is that due to the other agents' dispositions to act in compliance with coercion at *this* specific locus in the network, (v_4, v_5, v_{10}) will be reproduced – and so will d. Note that the value of d remains the same over the "generations", that is, after a reproductive cycle, where agent 4 was replaced by agent 6. Moreover, if submissive individuals leave the network and are replaced by other agents, the established power structure shapes their dispositions for acting, and they will have a tendency toward reproducing (v_{13}, v_{14}, v_{15}), as well. In such a scenario, the power structure as a whole persists through time, even if agents are replaced by other agents. Structural coercion is a social asymmetry that became institutionalized, and agent 4's slot can be taken by other agents because she and the other agents did NC before.

This is how cultural dominance or coercion reproduces structurally as a cultural niche. Furthermore, the amount of structural power of a cultural agent could be measured quantitatively in this way, given as the number of

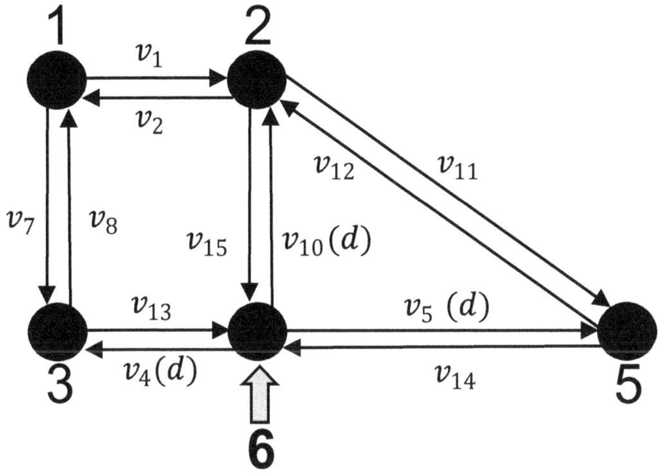

Figure 6.6 Another agent 6 enters the network. She will fill the power vacuum because agents 2, 3, and 5 were structurally adapted to agent 4's slot already.

other individuals that a dominant agent can compel to alter their behavior according to its directives.

6.5 Conclusion

While some facets of power have already been incorporated into CET by means of prestige-biased copying, the kind of power that poses a serious challenge to CET grants no room for personal preference among cultural variants. It is based on force or the threat thereof, namely cultural dominance or coercion. But not only is the lack of individual choice problematic for current CET modeling, but it is also the fact that dominance does not appear to be *copied* culturally. From a cultural evolutionary viewpoint, it does not seem to follow the logic of evolutionary fitness, according to which a trait's fitness increases as its frequency increases, since a dominant agent or an elite is able to make others do what is in their interest, not proliferate their cultural variants. Moreover, CET struggles to incorporate structural facets of power in general.

We suggested conceptualizing coercion and prestige as two distinct selective strategies. In the case of prestige bias, a cultural variant exhibited by a prestigious individual is preferentially copied. In the case of coercion, however, a cultural variant is enforced by a dominant individual or group via *different* cultural variants that are executed by subordinates. The model we propose puts cultural dominance in analogy to K-strategists in ecology due to its low reproduction rate but high success, while prestige resembles r-strategists as it has a high reproduction rate. This dichotomy between K-selected conservatism and r-selected innovation has profound implications for our understanding of cultural evolution. It posits that cultural traits are not merely passed down through generations but are actively selected based on the prevailing social and environmental conditions. These conditions are significantly influenced by power structures within societies, which can either hinder or promote the spread of cultural traits.

Furthermore, we have proposed that the reproduction of structural coercion can be viewed as a special form of *cultural niche construction*, where individuals do not just passively experience their cultural environments but actively shape them through their interactions. This process creates a self-reinforcing cycle, where the structures of power influence cultural evolution, which in turn shapes the very fabric of social structures.

It becomes increasingly clear that CET must expand its boundaries to encompass the multifaceted nature of power and its role in shaping human societies. Our conceptual integration of "power" into CET challenges the traditional focus on prestige and unveils the nuanced roles of coercion and dominance as pivotal forces in cultural propagation.

In conclusion, this chapter has endeavored to shed light on an often-overlooked aspect of power in cultural evolution, presenting a more nuanced perspective that recognizes the complexity of human societies. By integrating the concepts of r- and K-selection, as well as NC, we offer a more holistic view of how cultural traits are disseminated and how power dynamics shape the trajectory of cultural evolution.

Acknowledgments

We are deeply thankful to Martina Valković and Thomas A.C. Reydon for setting up this topical collection. Additionally, we thank them, as well as Grant Ramsey, Åsa Burman, and Azita Chellappoo, for their many helpful comments and constructive contributions to this chapter. Also, we are thankful for an anonymous reviewer's hint on the pseudo-scientific misuse of the r/K framework by figures like J. Philippe Rushton. The authors declare that there are no financial interests that have influenced this chapter.

Notes

1 Please note that the mathematical framework what we employ in Section 6.3 (the r/K Model) has historically been misused, notoriously by J. Philippe Rushton to support racist and biologically essentialist claims about human groups. Our use of the r/K model is entirely free of any such claims. It refers to the logic of cultural replication and structural persistence of behavioral patterns/variants – not to

any form of pseudo-biological or racial typology. We reject such interpretations unequivocally; see footnote 1 for more details.

2 We are aware that r/K selection theory has been misused in past work – most notoriously by J. Philippe Rushton – to support racist and biologically essentialist claims about human groups. Our use of the r/K framework is entirely free of any such claims and refers to the logic of cultural replication and structural persistence of behavioural patterns/variants, not to any form of biological or racial typology. We reject such interpretations unequivocally, they have nothing in common with what we want to communicate here. Besides, the r/K model is an ecological equation, which does not support any essentialism about groups of people like, for example "races", which in fact do not exist within the human species (see, e.g., Yudell et al., 2016).

3 For an interesting discussion of whether CET supports an individualistic or holistic view of culture and society, see Frank Hindriks (Chapter 4 this volume).

4 Our model does not currently account for large-scale infrastructural or institutional enforcement mechanisms (e.g., monopolies, states), which fall outside interpersonal coercion or prestige. This form of power may require additional modeling tools – possibly extending niche construction theory toward media infrastructures or economic ecosystems.

References

Aaby, B. & Ramsey, G. (2022): Three kinds of niche construction, *British Journal for the Philosophy of Science* 73: 351–372.

Arendt, H. (1958): *The Human Condition* (Second Edition), Chicago: University of Chicago Press.

Barkow, J.H. (1989): *Darwin, Sex, and Status: Biological Approaches to Mind and Culture*, Toronto: University of Toronto Press.

Boehm, J. (1999): *Hierarchy in the Forest: The Evolution of Egalitarian Behavior*, Cambridge, MA: Harvard University Press.

Boyd, R. & Richerson, P.J. (1985): *Culture and the Evolutionary Process*, Chicago: The University of Chicago Press.

Chellappoo, A. (2020). Rethinking prestige bias, *Synthese* 198: 8191–8212.

Cofnas, N. (2018): Power in cultural evolution and the spread of prosocial norms, *The Quarterly Review of Biology* 93: 297–318.

Currie, T. E., Campenni, M., Flitton, A., Njagi, T., Ontiri, E., Perret, C. & Walker, L. (2021): The cultural evolution and ecology of institutions, *Philosophical Transactions of the Royal Society B: Biological Sciences* 376(1828): 20200047.

Dahl, R. (1957): The concept of power, *Behavioral Science* 2: 201–215.

Durham, W. (1991): *Coevolution: Genes, Culture and Human Diversity*, Stanford: Stanford University Press.

Fracchia, J. & Lewontin, R. (2005): The price of metaphor, *History and Theory* 44: 14–29.

Geertz, C. (1973): Thick Description: Toward an Interpretive Theory of Culture, in: C. Geertz (Ed.): *The Interpretation of Cultures*, New York: Basic Books, pp. 3–30.

Haslanger, S. (2023): Systemic and structural injustice: Is there a difference?, *Philosophy* 98: 1–27.

Henrich, J. (2016): *The Secret of Our Success*, Princeton: Princeton University Press.

Henrich, J. & Gil-White, F. (2001): The evolution of prestige: Freely conferred deference as a mechanism for enhancing the benefits of cultural transmission, *Evolution of Human Behavior* 22: 165–196.

Laland, K.N. & O'Brien, M.J. (2011): Cultural niche construction: An introduction, *Biol Theory* 6: 191–202.
Lewens, T. (2015): *Cultural Evolution: Conceptual Challenges*, Oxford: Oxford University Press.
López, J. & Scott, J. (2000): *Social Structure*, Buckingham & Philadelphia: Open University Press.
Lukes, S. (1974): *Power: A Radical View*, London: Bloomsbury Publishing.
Mameli, M. (2008): Sociobiology, evolutionary psychology, and cultural evolution, in: Ruse, M. (Ed): *The Oxford Handbook of Philosophy of Biology*, Oxford: Oxford University Press.
Maner, J.K. (2017): Dominance and prestige: A tale of two hierarchies, *Current Directions in Psychological Science* 26: 526–531.
Maner, J.K. & Case, C.R. (2016): Dominance and prestige: Dual strategies for navigating social hierarchies, in: Olson, J. & Zanna, M.P. (Eds): *Advances in Experimental Social Psychology* (Vol. 54), San Diego, CA: Elsevier, pp. 129–180.
Mazur, A. & Booth, A. (1998): Testosterone and social dominance in men, *Behavioral & Brain Sciences* 21: 353–397.
Mesoudi, A. (2011): *Cultural Evolution: How Darwinian Theory Can Explain Human Culture and Synthesize the Social Sciences*, Chicago: The University of Chicago Press.
Mileva, V.R., Cowan, M.L., Cobey, K.D., Knowles, K.K. & Little, A.C. (2014): In the face of dominance: Self-perceived and other-perceived dominance are positively associated with facial-width-to-height ratio in men, *Personality and Individual Differences* 69: 115–118.
O'Connor, C. (2019). *The origins of unfairness: Social categories and cultural evolution*. Oxford University Press.
Pianka, E. (1970): On R and K selection, *American Naturalist* 104: 592–597.
Ramsey, G. (2013): Culture in humans and other animals, *Biology & Philosophy* 28: 457–479.
Ramsey, G. & De Block, A. (2017): Is cultural fitness hopelessly confused?, *British Journal for the Philosophy of Science* 68: 305–328.
Richerson, P.J. & Boyd, R. (2005): *Not by Genes Alone: How Culture Transformed Human Evolution*, Chicago: The University of Chicago Press.
Weber, M. (1921/1922): *Economy and Society*, Berkeley, Los Angeles & London: University of California Press.
Yudell, M., Roberts, D., DeSalle, R. & Tishkoff, S. (2016): Taking race out of human genetics, *Science* 351: 564–565.

7 Telic Power and Academic Excellence

Åsa Burman and Nemi Pelgrom

7.1 Purpose and Main Claims

The purpose of this chapter is both to introduce three categories of social power and apply those to the academic hiring process in Sweden. One part of our analysis is to *measure* the applicants' social power in this domain, and thus contribute to the empirical analysis of social power. This requires definitions of different concepts of social power, selecting the category that is most relevant for the analysis (telic power), and operationalizing telic power in a way that is useful in analyzing the empirical material. The empirical material consists of 1,341 external reviewer statements for tenured professorship positions in mathematics, philosophy, and political science at three main universities in Sweden from 2014 to 2024. In this way, we offer a new argument for the claim that the category of telic power is theoretically useful, a claim made by Burman (2023: 177).

We also argue that scholars in social ontology and cultural evolution should emphasize not only transparent forms of social power but also opaque forms of social power. Transparent forms of power cannot exist without being represented as such, while opaque forms of power can exist without being represented as such. One reason for paying attention to both types of power is the "opaque forms of social power objection" that we develop. Basically, that cultural evolutionary theory has difficulties or cannot accommodate opaque forms of social power since they are non-reproductive in a strong sense (the participants might not even know that the power exists).

The contribution is thus both theoretical and empirical: we clarify some relevant categories of social power and their relation to each other, apply one of these categories to the empirical material, with the result that women seem to have more positive telic power than men in this domain.

7.2 Power, Academic Excellence, and Gender Bias

Power creates and shapes the central parts of the social world. Who has power and the kind and extent of power involved create and shape social

DOI: 10.4324/9781003643845-9

phenomena like norms, practices, and institutions that deeply affect the life chances of individuals. At the same time, central parts of the social world, such as those norms, practices, and institutions, create and shape who has power, the different kinds of power there are, and what one can do with these kinds of power.

Consider the conceptual evolution of rights as an example. Some scholars describe a conceptual shift of special significance. At one point, some people were viewed as having *a claim* against the state (or king) not to take their property in return for money and military purposes. They had these claims, or rights, due to their gender and social and economic status. There was a radical shift from the idea that *some* people had rights in virtue of those traits to the idea of human rights – that *all* human beings have rights simply in virtue of being human (cf. Orend, 2002; Searle, 2010). This shift deeply affects the distribution of power in the social world and changes parts of that world since new institutions emerge to protect these rights (while some old institutions disappear). The conceptual evolution of human rights is also mainly about one specific form of social power known as "deontic power".[1]

In contemporary social ontology, institutional rights and obligations are referred to as deontic powers and come in two types: positive deontic powers are social or institutional rights, while negative deontic powers are social or institutional obligations. The basic idea is that once a person is assigned rights, she can do things that would not otherwise have been possible, such as vote in an election or live in a particular space until she decides to move (property rights). If she has an institutional obligation, she must do things that she would not otherwise have to do, such as filing taxes within a specific timeframe, or she must refrain from doing something that she would otherwise have done, such as moving into a certain space if it is not hers (an obligation to respect other people's property rights).

In the academic context, hiring committees have the right – a positive deontic power – to select a candidate for the position, and they have an obligation – a negative deontic power – to hire a candidate by following certain procedures. There is a parallel between academic hiring and the conceptual evolution of human rights that involves a shift from one basis of rights (social, economic, and gender) to another (being human). In academic hiring, the basis for being offered an academic position is not supposed to involve a particular kind of status, such as the gender one is, the university one comes from, or who one might know on the hiring committee. Rather, the hiring process aims to track and evaluate the candidate's skills or merit. The notion of meritocracy is the foundation of academic hiring processes in many parts of the world and a widely shared, important academic ideal.

But there are gaps between this ideal and practice due to different kinds of biases. For example, hiring decisions can be unduly influenced by what applicants' universities of origin, gender, and social connections are, rather than actual merit or skill. Some scholars argue that the standard of academic excellence is inherently male and biased against women (cf. Van den Brink & Benschop, 2012; Thornton, 2013). Or more specifically, that some criteria used for assessing quality favors men over women:

> we show that despite evaluators including diversity concerns in their search for talent during the first stages of the recruitment process, they end up deploying narrow criteria that tend to favour men in the crucial steps of the recruitment process, in which hiring outcomes are determined.
>
> (Orupabo & Mangset, 2022)

Thornton's study indicates that once more diffuse factors such as "potential" enters the discussion, it works against women:

> I was first struck by the malleability of merit from my experience of sitting on university selection committees. On one not atypical occasion, several well-credentialed women were passed over in the compilation of a short list in favour of an objectively less distinguished male candidate who was strongly supported by the chair of the committee. Although the chair conceded that the candidate's record was less than outstanding, what the favoured candidate had to make up for his shortcomings, the chair assured us, was "potential".
>
> (2013: 129)

Still other studies show that bases such as one's gender, or knowing someone on the committee, rather than skill or merit, play a large role in some decisions. Christine Wennerås and Agnes Wold showed that there was significant nepotism and gender bias in decisions made by Sweden's Medical Research Council (MRC), concluding that: "a female candidate had to be 2.5 times more productive than the average male applicant to receive the same competence score" (1997: 342). These decisions concerned who was to be offered the prestigious postdoc grants.

Our chapter is inspired by these studies. We offer a new case study related to meritocracy, gender, and academic excellence in Sweden by analyzing a large set of data through the lens of gender and different kinds of social power. The dataset consists of 1,341 external reviewer statements for tenured professorship positions in mathematics, philosophy, and political science at three universities in Sweden between 2014 and 2024.[2]

The reviewers are tasked with evaluating candidates against certain formal criteria that constitute a standard of excellence for research, teaching, and outreach, among other things: that is, an ideal of academic excellence.[3]

We ask what academic excellence is in practice by analyzing the data to uncover the details of the ideal of academic excellence in practice. We also examine the materials from a gender perspective to uncover whether there are gendered ideals and whether female and male candidates are evaluated differently. Finally, we ask whether female or male applicants have more power in this domain. To answer these questions, we use three notions of social power – deontic, telic, and artisanal – from contemporary social ontology. In the next section, we clarify these forms of social power.

7.3 Theoretical Framework: Three Categories of Social Power

The concept of power is central to understanding the creation, evolution, maintenance, and disruption of the social world. However, a wide variety of notions of power are on offer. We might think of examples in which the *physical power* of one person or group outweighs another's; this is power as brute force. Examples of successful performative acts (backed partly by brute force) establish a nation and a list of rights, such as the U.S. Constitution. This form of power is *social* in the sense that it exists partly because it is collectively believed to exist. The institutional rights and obligations referred to as deontic power are a form of social power (cf. Searle, 1995, on brute and social power and Barry Barnes, 1988, on the different bases of power).

Given the various understandings of power, it is crucial to clarify when we have the same concept of power in mind and might thus agree or disagree on the same topic, and when we are simply talking past each other because we have different concepts in mind.

John Rawls used the distinction between a *concept* of justice and a *conception* of justice in *A Theory of Justice* (1971). According to Rawls, people might have a particular concept of justice in common; for example, they might agree on the idea that justice means that everyone should get their due. However, what it means to get one's due is highly contested and a site of much disagreement. So, how to fill out what it means to get one's due is one's particular conception of justice, in Rawls's case, justice as fairness.

We also view this distinction between a concept and a conception as helpful in debates about power: when do we have different concepts of power in mind? When do we have the same concept but different conceptions of power in mind? For example, the difference between brute and social power may seem so stark that they are clearly about two different concepts of power. Recall their different conditions of existence and essential features – depending on brute force as opposed to collective belief to exist. We focus on the concept of social power in what follows, particularly on three different forms of social power.

Deontic powers are concerned with what we can demand from each other. In the academic hiring process, applicants can demand to have their submitted applications be considered, and the hiring committee is expected to offer reasons for their decision. Alternatively, consider the deontic aspects of the conceptual evolution of human rights. This is a shift from having positive deontic powers by virtue of one's social or economic status or gender to having rights by virtue of being human. We use John Searle's understanding of deontic power and the distinction between positive and negative deontic power, as described below:

> The first is where the agent is endowed with some new power, certification, authorization, entitlement, right, permission, or qualification granting the ability to do something he or she could not otherwise have done; and the second is where the agent is required, obligated, in duty bound, penalized, enjoined, or otherwise compelled to do something he or she would not otherwise have had to do – or, what amounts to the same thing, prevented from doing something that would otherwise have been doable. Roughly speaking, the two major categories are those of positive and negative powers.
>
> (1995: 100)

Note that this conception of social power refers to an *agent* having power and views power as an *ability* rather than saying, for example, that power only exists when it is exercised. It is also a *counterfactual* analysis of power in that it refers to what would otherwise have been doable. We return to our empirical case study and the academic hiring processes to illustrate this form of power: successful applicants obtain a new institutional right to a salary and resources from the university, which is a positive deontic power, while the university incurs a new institutional obligation to provide those resources, which is a negative deontic power. The reviewers have an institutional obligation to submit their reviews on time and in line with the instructions, and they have the institutional right to receive monetary compensation for their work. But how are external reviewers selected in the first place?

This question points to there being relevant forms of social power other than the deontic. The external reviewers are regarded as experts, and scientific expertise is partly about (being perceived as) living up to the ideal of academic excellence. That expectation and the standards of excellence point to a *teleological dimension* of the social world that concerns ideals or standards that others expect us to live up to and that we ourselves might want to fulfil. These standards give us reasons for action in the sense that we experience "a social ought". We might feel a push to comply with the standard, such as gender norms and being a good woman, or a pull toward it, such as trying to become a good or even excellent skier, tennis player, or academic. This teleological dimension can be contrasted with the deontic dimension discussed earlier, which concerns what we can demand of each other. Parallel to the deontic dimension and its related powers, there is a specific form of power related to the teleological dimension of the social world. In *Nonideal Social Ontology*, Burman identifies the category of telic power:

> TELIC POWER*: An agent A has telic power in a domain if and only if there exists an ideal such that agent A can be measured against it, and the distance perceived by other agents of A from the ideal affects A's ability to effect certain outcomes in that domain.
>
> (2023: 191)

This is a form of social power. One central feature, the existence of an *ideal,* is dependent on collective intentionality. Note also that it is *other agents' perceptions* that matter and give a certain agent an ability that they would not otherwise have had. This means that the ability is extrinsic rather than intrinsic to an agent. And it is a perception view of social power in the sense that it is other agents' perceptions that matter. The perceptions of someone's skills and their actual skills might diverge, and it is the former that matters for this form of social power. Similar to deontic power, it is an agent that has the power, and it is thus an *agent-oriented view of power*. Power is viewed as an *ability* rather than only existing if it is exercised, and it is a *counterfactual analysis* in the sense that it refers to what outcomes the agent could effect in virtue of having the telic power in question.

We return again to our empirical case study to illustrate this kind of power. In order to become an external reviewer, one needs to be perceived by other relevant agents as living up to the ideal of academic excellence. Once one is perceived as meeting this and other standards, one has a new institutional right to evaluate how well the various applicants meet the benchmarks of academic excellence. Note the relationship between these two different forms of social power; being perceived as meeting a standard (telic power) gives someone a positive deontic power. The relationship also obtains in the other direction; having the new deontic powers that come with being an external reviewer might affect other agents' perceptions and thus increase the external reviewer's "telic score", so to speak. In this way, these two forms of social power interact with and affect each other.

We can also imagine that some other suggested external reviewers were perceived as not living up to the ideal in whole or in part, and thus did not have the opportunity to offer to evaluate applicants. Consequently, we have positive telic power, or being perceived as living up to an ideal, and negative telic power, or being perceived as substandard in relation to an ideal. Burman defines positive and negative telic power in the following way:

POSITIVE TELIC POWER*: An agent A has positive telic power in a domain if and only if agent A is perceived as living up to the ideal; she is viewed by other agents as a good exemplar of the relevant kind, and this positively affects or enhances A's ability to effect certain outcomes in that domain.

NEGATIVE TELIC POWER*: An agent A has negative telic power in a domain if and only if agent A is perceived as not living up to the ideal; she is viewed by other agents as substandard or as a bad exemplar of the relevant kind, and this negatively affects or restricts A's ability to effect certain outcomes in that domain.

(2023: 191)

For example, agent A can have positive telic power in that she is viewed as an excellent academic, and this perception enhances her ability to effect certain outcomes, such as deciding on matters that concern a research group. Alternatively, she could be viewed as a substandard academic and have little say in the research group due to that perception.

We use the category of telic power in analyzing our dataset and answering our research questions: what is academic excellence in practice? Are there gendered ideals and evaluations? Moreover, we ask whether female or male applicants have more power in this domain. This category of power is particularly useful as a theoretical lens since its central components are *ideals* (an ideal of academic excellence), *other agents* (the external reviewers), and *their perceptions of the applicants* (reviewer statements comparing different candidates against the ideal).

However, some studies of academic hiring processes reveal a gap between the perceived and actual abilities of applicants due to different kinds of biases. One way of reformulating that divergence is to view it as a gap between an agent's telic and *artisanal power*, which is the third category of social power. It is particularly interesting since it does not depend on other agents' perceptions in the same way as telic power; rather, it is centered on actual abilities.

Charlotte Witt has recently developed an account of artisanal power (2025, manuscript) that relies on an Aristotelian framework, with central notions such as function, skill, and expertise. Witt's definition of artisanal power (manuscript, emphasis in original) runs as follows:

ARTISANAL POWER: Artisanal power is an agent's ability, capacity (or *dunamis*) to do something or to make something well according to local standards. Doing something well is both an instrumental power (i.e. the ability to achieve the goal) and a non-instrumental power (i.e. the ability to do it in the way it should be done around here).

Returning to our empirical case study, the hiring committee would presumably prefer there to be no gap or as narrow a gap as possible between applicants' perceived and actual abilities: that is, between their telic and artisanal powers.

The possibility of such a gap suggests that telic and artisanal power are very different notions of social power. The former is social in the sense that its existence conditions are ideals and the perceptions of other agents, while artisanal power relies on the intrinsic abilities of a person, say a carpenter, and that person's know-how of the relevant social practices and local standards of excellence in building a house in that local context. We thus appear to have not only two different conceptions of social powers but also two different concepts of social power, one that is perception dependent (telic) and one that depends on actual ability (artisanal).

To sum up this theoretical section, we focus on social power in contrast to brute power and distinguish between deontic and teleological dimensions of the social world. Within the deontic dimension, some social ontologists agree on the basic features of deontic power but fill out its basic features a bit differently. In those cases, we have different conceptions of deontic power.[4] Within the teleological dimension, we have identified not only two different conceptions of social power – telic and artisanal – but two different concepts of social power.

There is one explicit and one implicit premise in this section. The former is that all accounts of social power referred to are agent-centered accounts in the sense that it is agents that have the power in question, in contrast to accounts of power that do not locate social power within agents but rather emphasize either that structures have power or that power is more dispersed. Some accounts attempt to integrate the two (cf. Ledyaev, 2021, on trends in the conceptual analysis of power). The implicit is that the three accounts of social power are about transparent kinds of social power in contrast to opaque kinds of social power. Recent work by Muhammad Ali Khalidi helps clarify this distinction; he understands "opaque social kinds" in the following way: "The social kind and its instances can both exist without being conceived or represented as such (opaque)" (2025). The contrast is with "transparent social kinds", which require that a social kind be conceived or represented as such in order to exist. For example, a recession is an opaque social kind since it can exist without being conceived or represented as such, while money is an example of a transparent social kind since it cannot exist without being conceived or represented as such. *Transparent forms of social power*, then, can only exist if they are conceived or represented as such. By contrast, *opaque forms of social power* can exist without being conceived or represented as such.

The upshot is that most work in contemporary social ontology is about transparent forms of social power (cf. Burman, 2023, for a longer explanation) and thus leaves out a significant dimension of the social world (the opaque forms of social power).[5] Paying attention to the latter is crucial not only for social ontology (e.g., in fulfilling the aim stated by some social ontologists of providing a general theory of the social world) but also for cultural evolution. If we incorporate opaque forms of social power, it poses a new challenge to cultural evolutionary theory; the opaque forms of social power objection. We develop our objection in the next section.

7.4 The Opaque Forms of Social Power Objection

In this volume, two of the contributions on power (ours and the chapter by Baraghith and Jordan) follow the same trend of mostly omitting opaque forms of power. This chapter focuses on three forms of social power, but they are all transparent. Baraghith and Jordan note that there are many different understandings of power and settle on the following definition as a basis for their argument: if "power as a potential of A to make B do what is primarily in A's interest, or what B would not do in the absence of A, then A has power and B does not". This definition does not rule out opaque forms of power, but it is clear from the authors' subsequent discussion that they intend it as a transparent form of power in the sense that it can only exist if it is conceived or represented as such. The authors use an illustrative quote from the television series *Game of Thrones* to suggest that "power resides where men believe it resides". This understanding of power is also a transparent form of power in the sense that one of its existence conditions is a collective belief about its existence. Their argument is that cultural evolutionary theory can explain some forms of power, such as prestige, but either cannot or has difficulties explaining others, such as coercion and domination, since the latter do not rely on imitation. They then offer a novel solution to this problem by invoking interaction networks and K-selection.

By contrast, the chapter by Chellappoo invokes a structural account of power that we understand as an opaque form of social power. Chellappoo discusses the similar and well-known objection by Lewontin that cultural evolutionary theory cannot explain some important forms of power due to its commitment to methodological individualism; Chellappoo then strengthens it by using Sally Haslanger's account of gender and race (2000). In considering the strongest reply to this argument – that evolutionary game theory can indeed accommodate structural power and inequality of some sorts – Chellappoo counters by arguing that some structural phenomena and some inequality are macro-level phenomena that cannot be reduced to the micro-level. In short, both contributions start from key forms of power other than prestige. Baraghith and Jordan, as well as Chellappoo, then argue that cultural evolutionary theory either cannot capture or has difficulties capturing these forms of power. Baraghith and Jordan then try to save cultural evolutionary theory from the objection, but Chellappoo does not.

We want to offer a new argument against cultural evolutionary theory along similar lines. First, there is the distinction between opaque and transparent forms of social power and the definition of opacity: "The social kind and its instances can both exist without being conceived or represented as such (opaque)" (Khalidi, 2025). So, some forms of power can exist without being represented as existing. Second, recall Baraghith's and Jordan's analysis that cultural evolutionary theory cannot capture or, at least, has difficulties capturing non-reproductive forms of power, such as domination and coercion, because they do not rely on imitation.

Opaque forms of social power are also non-reproductive. Examples include a class or gender structure that the members of a society do not know exists (see Thomasson, 2003; Searle, 2010; Burman, 2023, on this debate). Amie Thomasson has shown that there can be types of social facts that are either ontologically or epistemically opaque. Given that opacity, these types of facts are also non-reproductive.

In sum, cultural evolutionary theory has difficulties or cannot accommodate transparent forms of power that are non-reproductive (coercion/domination). In addition, it is difficult or impossible for cultural evolutionary theory to accommodate opaque forms of social power. This is due to it being non-reproductive in *an even stronger sense* than domination and coercion, and thus not possible to imitate. Consider the contrast between social and economic class. In *Categories We Live By*, Ásta offers the historical example of a group of people with social standing who referred to the use of certain words and not others as upper class and hence conferring this property on only some people: "The Sloane Set decided what was U and non U (this is code for "upperclass"). Using certain words was U, others non-U. Using cloth napkins was U; paper napkins non-U. And so on" (2018: 50). Call this the Sloane Set example, which is an instance of a transparent social kind in which social classes must be represented or conceived as such to exist. Social class is a particularly nice fit with a form of power much discussed in cultural evolutionary theory: prestige, a reproductive form of power that can be imitated.

Contrast this to economic class understood as one's relationship to the means of production. It is an opaque social kind since it can exist without being conceived of or represented as such. Economic class might be either conceptually opaque (participants do not have access to the concept of economic class) or epistemically opaque (participants have access to the concept but do not think that their society is a class-based society). For both types of opacity, it means that one does not imitate "successful" individuals since one does not know that they are "successful" in this way. One person might have an upper-class social standing and be imitated due to prestige, but – unknown to the others – this person has recently lost all her property and wealth and has no way of regaining it.

This poses a specific challenge to the solution suggested by Baraghith and Jordan: can the interaction network and K-selection be used to resolve the opaque forms of social power objection? It also poses a general challenge for cultural evolutionary theory: can it accommodate opaque forms of social power that are non-reproductive in a strong sense?

We return to a transparent form of social power in the next section. More specifically, we use the category of telic power in analyzing the data and thus also show its theoretical usefulness.

7.5 The Empirical Study: External Reviewer Statements

We seek to discover the ideal or standard of academic excellence in a specific Swedish academic hiring context; that is, what are the features of the standard of academic excellence in practice? Furthermore, what is a good or even excellent academic? In other words, we try to uncover the standard, or ideal, of academic excellence. We also analyzed the empirical material from a gender perspective and used the concept of telic power: are there any differences in positive and negative telic powers for women and men? Is there any gender bias in these evaluations?

To narrow these large questions down to something manageable, we analyzed 1,341 external reviewer statements for academic positions within Sweden. This material is public acts in Sweden. We reached out to three universities, Stockholm University, Lund University, and Linnaeus University, and asked for all the external reviewer statements for academic positions in political science, mathematics, and philosophy during 2014–2024. We chose these fields because we wanted to have one example from the social sciences, one from the natural sciences, and one from the humanities.

It will be helpful to clarify some things about the hiring practices in Swedish academia. There are usually two external reviewers for each academic position. They are tasked with evaluating candidates against specific formal criteria about research, teaching, administrative leadership, and outreach, among other things. The external reviewers read the submitted materials from the applicants, such as publications, CVs, teaching evaluations, letters of recommendation, etc. They make a shortlist and then conduct a more extensive review of the shortlisted applicants. The hiring committee, which is an internal committee at each university, uses these external reviewer statements, lectures by the shortlisted applicants, and interviews with the applicants to decide who is to be offered the academic position. The external reviewer statements play a key role in the hiring decisions – it is seldom that the internal committee overturns their ranking. It is also important to note that there is no formal requirement that applicants from an underrepresented gender must be on the shortlist. However, if there are two applicants with the same merits, then the hiring committee can choose the underrepresented gender.[6]

7.6 Methodology

For the data collection phase, we requested all external review statements for mathematics, political science, and philosophy from 2014 to 2024 from Lund University, Linnaeus University, and Stockholm University. We received 1,341 external reviewer statements, of which 23 had to be transcribed by us due to being in non-digital form. The received documents were of different formats, with varying information in them. So, we extracted the information manually to ensure that we gathered the correct information from each.

For the data cleaning phase, we extracted the qualitative description of the applicants, along with their names and genders. The gender was assigned according to how they were mentioned in the qualitative texts. Furthermore, we collected the names, affiliations, countries of affiliation, and genders of each of the experts. The gender was assigned from their names and pictures. All texts that discussed several candidates at the same time, as opposed to consecutively, were excluded due to them explicitly comparing candidates with each other, making them difficult to compare with the same methods as the other texts.

For the data analysis phase, all numbers were collected through Excel formulas on the qualitative texts and their annotations. Most were collected by formulas that counted keywords or annotated words. To gather keywords that were relevant to this dataset, we created a list of keywords that we expected to find. Our expectations were based on our knowledge of the formal criteria used in job advertisements, the Swedish academic context, and we also read random samples of the collected texts. Restricting our list to words that appeared in at least 5% of all texts gave us a total of 49 words (Table 7.1). The keywords were on topics related to the ideal researcher or the ideal academic, and academic excellence.

One complication is that some texts are written in English, and some are in Swedish. To account for this, some of the numbers we present in the results section are combinations of formulas using an English term and a corresponding Swedish translation of this term. Both authors are fluent in Swedish and English, and we only combined the terms if they were judged to be equal by both of us. For words that might be used in several different ways, we manually checked at least 25% of all instances of these words to avoid collecting numbers that are misrepresentative. One such example is the word "doktorand" which means PhD student, may be used to refer to what the candidate was doing as a PhD student, or that they are teaching a course at that level. However, all of the instances in this check were referring to the amount of supervision of PhD students the candidate had provided, which is what we were interested in. Furthermore, it was interesting to study the keywords in the external reviewer statements outside of how they are used in the specific evaluations of a particular candidate. Some words will be used in either a positive or negative light (e.g., "impressive" and "not impressive"). However, we will only note the number of times these words have been used, rather than whether they have been used in a positive or negative light, due to our interest in how these words are used to describe an ideal candidate and how an ideal candidate is interpreted. Describing a candidate as "impressive" or "not impressive" both indicate that being impressive is a measure that is used and that it is something to value in the context. A final note on the data analysis is that we decided which words to include in the analysis before we got access to the numbers connected to the words. When categorizing, we paid more attention to how the words are used in this context rather than the words as potential general indications of ideals.

Concerning error sources, it is important to note that all qualitative text was manually copied and annotated into the working document. The reason for this was the variety of formats used in the source documents, thus making it difficult to automate such an extraction. It is possible that some data have been pasted in the wrong format or missed. Another possible source of error is that, due to the different formats of the documents, some are dividing the text into sections with titles such as "Pedagogical skills". Such instances may affect the counts of the words used since the word count is on all of the extracted text. Another possible source of error is that we transcribed some of the documents, and some of these may contain the wrong spelling of some words since we have not read through all of the text in the dataset. We have skimmed everything, so we know that there are no extreme outliers, but it is possible that some words are missed in the count.

A few documents were excluded because they contained text that was not possible to extract without being edited, and we decided to prioritize keeping the texts as they were without editing them at all. Some experts used only gender-neutral text, and these texts were included but were not annotated with any gender. So they were counted toward our identification of telic power or the standard of the ideal researcher, but were not part of the analysis of gender.

7.7 The Results of the Empirical Study

In uncovering the ideal of academic excellence, we classified the most prevalent words into four categories – internal, productivity, external, and comparative – and used the first three categories to uncover the ideal used in practice. Roughly, *internal* concerns activities have to do with basic and applied research to increase knowledge, while *external* concerns activities, such as interactions with others as part of the goal, include outreach and teaching students. *Productivity* concerns how the output in the different areas is measured. We summarize the results in Table 7.1.

The most common words used in the category "Internal" are these: Active, Impressive, Strong, Excellent, High, Skilled, Well, Scientific, Promising, Clear, Independent, Ability, Handled, Quality, Focus, and Very. In this category, we have both properties, such as being scientific, focused, and independent, and to what extent someone is focused, such as high, very, and well. The five most common words in this category are as follows: very, focused, scientific, well, and high. All are used more than 400 times.

The second category is "Productivity". In this category, we have qualitative evaluations, such as to what extent something is relevant or interesting; quantitative statements, such as to the extent someone has an extensive publication record; and again, an assessment of what extent someone is something, such as top in their field. The most common words in this category are Quantitative, Extensive, Top, Developed, Substantive, Productive,

Table 7.1 Classification of the most prevalent words

Ideal internal		*Ideal productivity*		*Ideal external*	
Active/aktiv	220	Quantitative	127	Teach	873
Impressive	150	Extensive	207	Supervis	555
Strong	231	Top	232	PhD Students	99
Excellent	71	Developed	137	Undergraduate	200
High	403	Substant	96	Group	305
Skicklig	109	Productive	80	Organized	78
Well	409	Peer-reviewed	226	Outreach	100
Scientific/Vetenskaplig	409	Important	136	Public	200
Promising	107	Interesting	141	Co-authored	218
Clear	292	Internation	701	Collabo/Samarbet	252
Independen	66	Releva	258	Grant_	128
Ability	213	Experience	789	Funding	200
Handled	112	Stor	280	External funding	69
Quality	233	Major	145	Leader	166
Focus/Fokus	464	Significa	146	Samverka	79
Very/Mycket	519			Editor	86
				Pedagogi	461
				Administra	379

Peer-reviewed, Important, Interesting, International, Relevant, Experience, Large (Stor), Major, and Significant. The words "international" and "experience" are the two most prevalent categories, with more than 700 uses each.

The third category is "External". In this category, the most common words are as follows: Teach, Supervision, PhD Students, Undergraduate, Group, Organized, Outreach, Public, Co-authored, Collaboration, Grant, Funding, External funding, Leader, Cooperate, Editor, Pedagogic, and Administration. The words "teach", "supervision", and "pedagogic" stand out with more than 400 uses each.

From the perspective of our quantitative analysis, academic excellence is about being engaged in teaching (873) and supervision (555), being experienced (789), engaged internationally (701), and being focused (464), pedagogical (461), and scientific (409). Furthermore, words are commonly used to link someone to how well that person fulfills a standard, such as very (519) and well (409). In short, relative to this dataset, the standard or ideal of being an excellent academic consists of *being engaged in teaching, being experienced, and being engaged internationally.*[7]

How do external reviewers perceive the candidates according to gender? We found that women are significantly overrepresented (i.e., a difference of more than 15%) with respect to the following words: "cooperate", "much", and "focus". Meanwhile, men are significantly overrepresented (i.e., a difference of more than 15%) with respect to "PhD students". It is relevant to note that both genders receive a proportionate amount of text written about them by the external reviewers.

We also found that women overestimate themselves just as much as men in the sense that an equal proportion of women are excluded from being evaluated by external reviewers due to being viewed as not meeting the formal hiring criteria. This result is conditional on the assessment that the candidates do not meet the formal criteria, such as having a PhD in the relevant subject, being correct. We summarize the results in Table 7.2.

Table 7.2 Summary of our findings

	Count	*Percentage (%)*
Men among applicants	910	69
Women among applicants	403	31
Men shortlisted	245	64
Women shortlisted	135	36
Men deemed unqualified	69	68
Women deemed unqualified	32	32
Text about men	294,671	67
Text about women	141,921	33

Another result concerning gender is that only about a third (69% male, 31% female) of all applicants are women. But women are overrepresented (64% male, 36% female) on shortlists. This is a surprising result given the results of other research on gender bias in academia (Wennerås & Wold, 1997). We discuss this in greater detail in the next section.

7.8 Discussion

The result that women are overrepresented on shortlists might seem surprising at first, given other research about gender bias in academia. Wennerås and Wold, for instance, found a significant gender bias, and other studies also show similar results (Van den Brink & Benschop, 2012; Thornton, 2013; Orupabo & Mangset, 2022). In response, one might argue that even though women are overrepresented on shortlists, this does not mean that there is no bias against women (or men, for that matter) in the academic hiring processes in the dataset. For example, it might still be the case that women's (or men's) merits are still not appraised accurately[8] or that a greater proportion of men are chosen from shortlists for positions, so that even though women are overrepresented on shortlists, relatively few earn a position in the end.

To analyze this result and respond to these concerns, the three categories of social power are useful. Recall that telic power is perception dependent, while artisanal power is not. There can thus be a gap between an agent's telic and artisanal power. Consequently, the results are consistent with there still being gender bias in the academic hiring process. What has been measured is the applicants' telic power rather than either their artisanal power or the gap between the two forms.[9]

In fact, telic power is especially apt as a theoretical tool to analyze the empirical material. The external reviewers evaluated the applicants against specific criteria constituting a standard of excellence for research, teaching, administrative leadership, outreach, and the like. The definition of telic power takes shared *ideals*, *other agents*, and *perceptions* to be central components of this particular kind of social power. We uncover an ideal being used by the external reviewers (the other agents) in practice. We also analyze how the external reviewers perceive the different candidates according to this standard.

Contrast this to deontic power and artisanal power, which would be less useful as theoretical lenses (but still central in other respects). Deontic powers do not include shared ideals and the evaluative component that moves by degrees (fulfilling the standard to a greater or lesser extent). Rather, it is binary (one has certain institutional rights in virtue of one's employment by a university or one does not due to not being employed by that university). Deontic powers are still important in understanding other aspects of the academic hiring process, such as who can serve on internal hiring committees (one requirement is having a PhD or the equivalent) or which applicants gain new deontic powers in the end due to being offered a position. Likewise, artisanal power is key to understanding central aspects of the process; for example, external reviewers' statements are used to track the applicants' actual abilities. But it is their perceptions of the applicants' merits that are reported, and thus, telic power is the more useful notion for these purposes. And as previously noted, bias in the academic hiring system can be reformulated as a gap between an applicant's telic and artisanal power. This brings us to the following question: are there any differences in telic power between female and male applicants? Women seem to have more positive telic power than men in this domain. This result is to be understood in terms of the quantitative analysis of the external reviewers' written reports, where overrepresentation on shortlists means that female applicants (as a group) were perceived as meeting the standard of academic excellence to a greater extent than the male applicants (as a group). So, women have more positive telic power than men *in relation to the external reviewers' written reports* (which play a major role in the hiring decision).

However, this result does not include which candidates were ranked *highest* on the shortlist and *who was offered* the position at the final stage, when the external reviewers met with the internal hiring committee at a given university. If men turned out to be overrepresented with respect to being offered a position, then women would not have more positive telic power than men in this domain, taking into account not only the external reviewers' written reports but also the final – and most important – stage of the hiring process.

Another result worth discussing is that only a third of the applicants are women. Two of the fields are mathematics and philosophy, which are heavily male-dominated. Even if the women who applied have more telic power than the men who applied in those domains, we might need other categories of power to answer the related question of why it is the case that so few women applied relative to men. Here, none of the three forms of social power is helpful. Telic power concerns being measured against an ideal, but we want to know why so few women were measured in the first place. Deontic power is about institutional rights and obligations, such as meeting the eligibility requirements. One (deontic) explanation is that there are fewer women with PhDs in philosophy and mathematics than men, but that merely moves the question back one step: why is this the case? We might thus need to turn our attention to other forms of social power to have the categories needed to explain this phenomenon. The opaque forms of social power seem particularly interesting because they can exist without being represented as such. For example, one might need to use categories such as structural power and social structures to explain this. Consequently, it is crucial to be clear about what we mean by social power by distinguishing between different forms of social power and then choosing the right tool – in this case, telic power – for one's purposes.

7.9 Conclusion

This chapter contributes to social ontology and cultural evolution in three related ways. The first conclusion is empirical. Our study of 1341 external reviewer statements for academic positions in Sweden shows that the ideal, or standard, of academic excellence in this particular domain is *being engaged in teaching, being experienced, and being engaged internationally*. More specifically, from the perspective of our quantitative analysis, academic excellence in this domain is about being engaged in teaching (873) and supervision (555), being experienced (789), engaged internationally (701), and being focused (464), pedagogical (461), and scientific (409).

This standard was uncovered by summarizing the most frequent words used in evaluating candidates for an academic position. That is, the words used in this academic practice of evaluating candidates, rather than the formal criteria in the position advertisement. Furthermore, we found that female and male candidates had equal amounts of text written about them. We also found that women were very overrepresented concerning frequency in the following words used to describe them, namely "focused", "very", and "collaboration", while men were very overrepresented with respect to the word "PhD student". Another interesting result is that women are overrepresented on the shortlists. However, women comprised only about one-third of the total applicants. This suggests that women have positive telic power to a larger extent than men in this particular domain. In other words, the female applicants are perceived as an ideal academic, or meeting the standard of academic excellence to a larger extent than men by the external reviewers.

The second conclusion is theoretical. The result that women seem to have positive telic power to a larger extent than men in this particular domain supports the claim that telic power is theoretically useful (Burman, 2023). Using the definition of telic power in analyzing the empirical dataset and uncovering a standard of academic excellence is a way of operationalizing telic power. Another is that it can be used to measure the telic power of different agents in a particular domain. More generally, our aim was to show that concepts from social ontology, such as telic power and artisanal power, are useful in analyzing particular academic hiring practices and the potential gap between an applicant's perceived and actual abilities. The relevance of introducing these categories in the debate about academic hiring practices is both to clarify what exactly we are talking about (the different forms of power) and to offer a helpful way of reformulating the role of bias discovered in academic hiring practices, as a gap between an agent's telic and artisanal power. The rationale for choosing telic power as the main theoretical lens is that other agents' perceptions of how well the applicants fulfill the academic ideal (all key components of telic power), more or less, are supposed to track actual academic skills, which are a major factor in the final job decisions. Put succinctly, we have measured social power. More specifically, we have measured the telic power of female and male applicants in a particular academic context. This suggests that telic power can be used as a theoretical tool in other empirical studies, too.

This brings us to the third conclusion, which is methodological. Power is a central social concept relevant to social ontology and cultural evolution. We might have different and more or less precise ideas in mind when speaking about power and using it as a theoretical tool. John Rawls used the concept–conception distinction in his *A Theory of Justice* (1971). We take this distinction to be helpful in debates about power as well. When do we have different concepts of power in mind? For instance, the difference between brute and social power is about different concepts of power due to their different existence conditions and essential features. When do we have the same concept but different conceptions of power in mind? For example, many social ontologists share an understanding of what deontic powers are. So, they have different conceptions of deontic power.

The concept–conception distinction is thus useful for clarifying whether one has a genuine or merely apparent disagreement with another researcher due to discussing the same (or a different) concept. For example, some of the debates about power in this volume seem to be about two different concepts of power, such as agent-centered and not-agent-centered views on power, or between brute power and social power, and the authors might thus talk past each other.

The two categories of power of most significance in this chapter are telic and artisanal power. Are they different concepts of power or different conceptions? They both concern the teleological dimension of the social world and thus share some features, such as ideals or standards, which are essential components of the respective definitions. Another essential component is that they both refer to an agent's ability. Power is thus a dispositional concept in both accounts. Nevertheless, there is a key difference between them: telic power refers to perceived ability, while artisanal power refers to actual ability. They are also social in different ways; telic power refers to ideals and perceptions of other agents, while artisanal power refers to local standards of excellence. Within the sphere of social power, then, telic power and artisanal power seem to be two different concepts of power.

In general, we hope to have convinced the reader of the importance of clearly defining and distinguishing between different types of power, that telic power is theoretically useful, and the importance of paying attention to opaque forms of social power.

Acknowledgments

The authors wish to thank Azita Chellappoo, Joshua Rust, Karim Baraghith, Martina Valković, and Thomas A.C. Reydon for helpful comments. An earlier version of this chapter was presented at a workshop on Social Power at the University of Neuchâtel in 2025 and at the *Social Ontology 2025 Conference* in Dublin. We thank the participants at those occasions for their constructive comments.

The authors declare that there are no financial interests that have influenced this manuscript.

Notes

1 Human rights are usually (and plausibly) viewed as moral claims too.

2 A typical example of the formal criteria is this advertisement for a lectureship in theoretical philosophy at Stockholm University: Doctorate in philosophy or equivalent scientific competence.

Documented experience of teaching at university level, including course development and examination. Experience of research in the theory of science and publication in scientific journals.

Ability to teach and supervise students in Swedish and English.

Good collaboration skills and ability to contribute to a positive work environment.

Merits: Experience of interdisciplinary collaboration, Experience of research funding, and Experience of leading research projects.

3 In general, for positions in Sweden, research and teaching are the two main categories of evaluation. For assistant and associate professorships ("lektorat"), they should be given equal weight. Administration, leadership, and outreach are often other things being evaluated, although not given the same weight as the two main categories. Research usually includes statements like having strong research record, including publications, external funding, and participation in research projects. Teaching concerns having demonstrated teaching experience and pedagogical skills, often evaluated through a teaching portfolio or trial lectures if an applicant is shortlisted. The qualification is doctoral degree (PhD) or equivalent research expertise. The university can also add *merits* beyond this basic qualification, for example, experiences of interdisciplinary work, and having led a research project.

4 For example, Ásta's conferralist account of deontic power is similar to Johan Brännmark's account of deontic power, explicated in terms of Hohfeldian incidents (Brännmark 2019). Both accounts refer to deontic powers as formal and informal rights and obligations, and state similar conditions of existence in the sense that deontic powers depend on the perceptions and behaviors of other agents to exist. Thus, these two accounts are about two different conceptions – rather than different concepts – of deontic power.

5 For a diagnosis of why this is the case and a remedy in the form of an account of social power that includes both transparent and opaque forms of social power, see Burman (2023).

6 All three universities have signed the Declaration of Research Assessment (DORA), so there is no difference between them in this respect. The general recommendation is as follows: "Do not use journal-based metrics, such as Journal Impact Factors,

as a surrogate measure of the quality of individual research articles, to assess an individual scientist's contributions, or in hiring, promotion, or funding decisions". https://sfdora.org/read/

7 It is also interesting to consider what does not stand out, for example, funding and standing or status in the field. We found that the word "administration" stands out with more than double the amount of uses compared to the word "funding". It is also noteworthy that the ability to get funding (represented by words like "funding", "grant", and "external funding") and "standing" in the field (represented by words like being an editor) is mentioned 200 times or less each. It might be the case that this is due to some of the applicants being less senior and thus not having gained this standing yet, or the funds associated with it.

8 We are grateful to Azita Chellappoo for posing this objection.

9 One might think that it is rather a gap between an applicant's telic power based on the external reviewers' statements and that applicant's telic power based on the publication record (depending on how one views the mix of abilities measured in the peer-review process).

References

Ásta (2018): *Categories We Live By*, New York: Oxford University Press.

Barnes, B. (1988): *The Nature of Power*, Champaign: University of Illinois Press.

Brännmark, J. (2019): Contested institutional facts, *Erkenntnis* 84: 1047–1064.

Burman, Å. (2023): *Nonideal Social Ontology*, New York: Oxford University Press.

Haslanger, S. (2000): Gender and race: (What) are they? (What) do we want them to be?, *Noûs* 34: 31–55.

Khalidi, M.A. (2025): Transparency, collective intentionality, and social power, *Philosophy of the Social Sciences* 55(6).

Ledyaev, V. (2021): Conceptual analysis of power: Basic trends, *Journal of Political Power* 14: 72–84.

Orend, B. (2002): *Human Rights: Concept and Context*, Peterborough: Broadview Press.

Orupabo, J. & Mangset, M. (2022): Promoting diversity but striving for excellence: Opening the 'black box' of academic hiring, *Sociology* 56: 316–332.

Rawls, J. (1971): *A Theory of Justice*, Cambridge, MA: The Belknap Press of Harvard University Press.

Searle, J. (1995): *The Construction of Social Reality*, New York: The Free Press.

Searle, J. (2010): *Making the Social World*, Oxford: Oxford University Press.

Thomasson, A. (2003): Foundations for a social ontology, *ProtoSociology* 18–19: 269–290.

Thornton, M. (2013): The mirage of merit: Reconstituting the 'ideal academic', *Australian Feminist Studies* 28: 127–143.

Van den Brink, M. & Benschop, Y. (2012): Gender practices in the construction of academic excellence: Sheep with five legs, *Organization* 19: 507–524.

Wennerås, C. & Wold, A. (1997): Nepotism and sexism in peer-review, *Nature* 387: 341–343.

Witt, C. (2023): *Social Goodness*, New York: Oxford University Press.

Witt, C. (2025) manuscript: *Social norms and social powers.*

Part III

Interactions and Collectives

8 The Evolution of Coordination and Common Ground

Bart Geurts

8.1 Introduction

Whether she is having dinner with her family, shopping for groceries, or strolling in the park with a friend, Betty constantly needs to coordinate her actions with others', and doing so requires a fair amount of sharing. Take the first case. In addition to food sharing, a family dinner also requires shared knowledge about who is who and what they are doing together, shared opinions about what is appropriate behavior at a family dinner and what is not, and, of course, it requires a shared language. This and much more is common ground between Betty and her family, and without it, having dinner *together* would be hard to imagine. More generally, everything we do with others requires common ground.

Common ground is one of a family of theoretical concepts that have been developed since the 1960s, which includes "common knowledge" (Vanderschraaf & Sillari, 2023; Yalcin, 2024), "mutual knowledge" (Schiffer, 1972; Clark & Marshall, 1981), "mutual belief" (Bach & Harnish, 1979; Stalnaker, 2014), "mutual acceptance" (Stalnaker, 2014), "joint commitment" (Clark, 1996, 2006; Gilbert, 2014), and "mutual commitment" (Geurts, 2019, 2024). All these notions have been argued to involve higher-order sharing. For example, if it is mutually known between Betty and her friend that they are strolling in the park, then at a minimum:

(1) They share the knowledge that they are strolling in the park.
(2) They share the knowledge that (1) holds.

Ditto, mutatis mutandis, for mutual belief, mutual commitment, and the rest. Following the lead of Clark (1996) and others, this chapter adopts an inclusive use of the term "common ground", in that mutual belief, joint commitment, etc., are treated as modes of common ground, all of which involve higher-order sharing (Section 8.2). Common ground, thus understood, is a family of structures with a highly diverse membership.

DOI: 10.4324/9781003643845-11

Modes of common ground are stock-in-trade in the social sciences and the humanities, where they have been enlisted by theories of communication, conventions, institutions, norms, collective action, joint attention, and so on (cf. the chapters by Slors and Stoelhorst in this volume). Thus, we will be exploring a very general perspective on human sociality and social ontologies.

By all accounts, achieving common ground is cognitively demanding and therefore likely to be a privilege of humans. This raises the question of how human sociality could have evolved to become essentially dependent on common ground. Given that the chief purpose of common ground is that it enables us to coordinate our activities, this is part of a larger issue: the evolution of distinctively human forms of coordinated action, of which human communication is a notable special case.

The first and thus far only one to explore an evolutionary perspective on common ground was Armstrong (2018, 2025), who convincingly argued that, by construing common ground as higher-order sharing, we fail to account for two observations: first, that many non-human animals apparently coordinate their activities in human-like ways, but without common ground; and second, that action coordination in humans doesn't always rely on common ground, either (Section 8.3). In order to accommodate these observations, Armstrong defined a weaker notion of common ground, which does not require higher-order sharing and is therefore easier to achieve than conventional common ground. Although I support Armstrong's project, I am concerned that his notion of common ground still excludes a great many life forms that engage in coordinated behaviors. This is not to devalue Armstrong-style common ground at all, but it is to say that, in addition, an even more general notion is needed. In that spirit, I define "shared ground" as a minimal form of sharing that subsumes the standard modes of conventional common ground (i.e., higher-order sharing) as well as Armstrong-style common ground (Section 8.4).

In the second half of the chapter, I first present a number of cases in which primitive forms of coordinated action are enabled by shared ground but neither by higher-order sharing nor by Armstrong-style common ground (Sections 8.5–8.8). The chapter ends with a discussion of the relationship between shared ground and higher-order sharing, with special emphasis on normative modes of common ground (Sections 8.9–8.11).

This chapter presents a framework that accommodates common ground as an advanced form of sharing, which was preceded by and is embedded in an abundant variety of more primitive forms. A fully fledged account of how the former emerged out of the latter is not on the agenda. That would require a general theory of the evolution of human sociality, which is too tall an order for a single chapter. However, as we go along, we will touch on various aspects of the transition from shared to common ground.

8.2 Common Ground as Higher-Order Sharing

Common ground is often viewed as a container holding information shared between agents, which is typically modeled by a set of propositions that are "in" the common ground (Stalnaker, 1978). The problem with this view is not that it is wrong, but that it is so abstract that it doesn't explain much. In particular, it fails to capture a distinction that is widely agreed to be essential, namely between information that is common ground and information that is merely shared. Consider the following two cases. In one, both Betty and Wilma happen to believe that Stalin was gay. In another case, they discussed the matter and came to agree that, yes, Stalin was gay. The difference is evident and important, but mere sets of propositions will not capture it. More structure is needed.

Enter higher-order sharing. Suppose that Betty and Wilma not only share the belief that Stalin was gay, but also share the belief that they share that belief. Then Betty's and Wilma's beliefs are tied together by second-order beliefs about each other's beliefs; they *mutually* believe that Stalin was gay. Second-order shared belief is the weakest form of mutual belief, and it is often held that more than two orders of sharing are required for mutual belief, though a small number of orders will do. Thus, Bach & Harnish (1979: 5) define mutual belief as third-order belief sharing. On their account, A and B mutually believe that φ iff:

(1) A and B believe φ.
(2) A and B believe (1).
(3) A and B believe (2).

Others define mutual belief as an infinite progression of higher-order belief sharing:

(1) A and B believe φ.
(2) A and B believe (1).
(3) A and B believe (2), and so on.

For our current purposes, these differences don't matter. All we need here is that, by all accounts, mutual belief is n-order belief sharing, for some $n > 1$. This suffices to set mutual belief apart from merely shared belief (i.e., first-order belief sharing), and in a way that matters.

If common ground is mutual belief, it is a higher-order sharing of beliefs. On this view, common ground is not just a set of propositions, but a structure of beliefs, and therefore it is, strictly speaking, false to say of any proposition φ that it is *in* the common ground, as the container view has it. It is more accurate to say, instead, that φ *is* common ground; it has the property of being mutually believed.

The view that common ground is mutual belief is mentalist in the sense that it treats common ground as an interpersonal structure of psychological states. Different members of the mentalist family use different psychological states as the building blocks of common ground, while keeping the structure of mutual belief. For example, Stalnaker (2014) defines common ground as mutual acceptance (with infinitely many orders of acceptance sharing). Acceptance, according to Stalnaker, is weaker than belief, which makes it possible for a speaker to accept a proposition, for example, for the sake of an argument, without believing that it is true. At the same time, Stalnaker holds that belief and mutual belief "are the default settings for acceptance for the purpose of the conversation (and for common ground)" (p. 45). That is to say, while acceptance is weaker than belief, it defaults to belief, ceteris paribus, and by the same token, common ground defaults to mutual belief, ceteris paribus. Hence, Stalnaker effectively proposes a pluralist account in which mutual acceptance and mutual belief coexist, and the latter is the "default setting" for common ground. My view on common ground is pluralist, too, but it is more inclusive than Stalnaker's.

Belief is a stackable attribute: we have beliefs about beliefs, beliefs about beliefs about beliefs, and so on. Stackability enables groups of believers to have mutual beliefs, which helps them coordinate their activities. The same holds for other belief-like states, which aggregate in forms of higher-order sharing with the same structure as mutual belief: mutual acceptance, mutual knowledge, and so on.

Like belief and its ilk, some *non*-psychological attributes are stackable, too. For example, "having reason to believe", which is a key concept in Lewis's (1969) monograph on convention, is a stackable attribute, and Lewis's theory includes an account of how individuals' reasons to believe aggregate into mutual reasons to believe that φ:

(1) A and B have reason to believe φ.
(2) A and B have reason to believe (1).
(3) A and B have reason to believe (2), and so on.

Reasons to believe are not beliefs: one can have reason to believe φ without believing φ, and vice versa. Rather, reasons to believe are normative attributes, which our beliefs *ought* to conform to, though they may fail to do so.

When viewed as a structure of reasons to believe, common ground is a normative structure. Another, rather different normative way of construing common ground is in terms of commitment:

(1) A and B are committed to the truth of φ.
(2) A and B are committed to the truth of (1).
(3) A and B are committed to the truth of (2), and so on.

Mutual commitment is at the heart of my approach to pragmatics (Geurts, 2019, 2024), in which it does much of the explanatory work done by mutual belief in other frameworks. I will have more to say about commitments later on, but for now it suffices to note that mutual commitment and mutual reason to believe provide us with non-mentalist, normative ways of construing common ground.

Despite the fact that there is no consensus on what common ground is and a variety of views seem viable, *all* define common ground as higher-order sharing of some attribute or other. This agreement across a range of otherwise disparate accounts may seem remarkable, but considering the kind of creatures that humans are, it isn't that surprising. We are an extraordinarily social species with extraordinary cogitative powers, which enable us to relate to each other in complex ways, including ways that are responsive to ways we relate to one another. Iterative patterns are to be expected, and it shouldn't come as a surprise that the same iterative structure has become a feature of sundry forms of sharing that enable us to coordinate our actions. In a nutshell, this is the inclusive, pluralist view on common ground as higher-order sharing that I proposed in Geurts (2024).

On the pluralist view, common ground isn't one structure, but rather a family of structures. Given that beliefs are in the psychological domain, commitments are in the social domain, and reasons to believe are yet somewhere else, mutual beliefs, mutual commitments, and mutual reasons to believe are in disparate domains, too. They are not parts of one structure; rather, common ground is a structure that emerges in disparate domains. In this sense, we might say that mutual beliefs, mutual commitments, and mutual reasons to believe are "modes" of common ground. When, in the following, I say that something is common ground tout court, this may be understood as implying that it is common ground in one or more modes, and usually the modes I have in mind are mutual commitment and mutual belief.

Regardless of how the distinction between first- and higher-order sharing is worked out, there is a sharp line between the two conditions, which accords with the general feeling that "mere sharing" is not good enough to support human social interactions, in general, and their communicative exchanges, in particular. Despite this sentiment, Armstrong (2018, 2025) argues, in effect, that there are forms of first-order sharing that enable a wide range of social interactions, not only in humans but also in some other species. I agree, and will extend Armstrong's line of thinking across and beyond the animal kingdom.

8.3 Coordination Without Higher-Order Sharing

Armstrong argues that current mentalist theories of common ground are defective. His critique specifically targets meta-representations, which in our terminology are second-order mental representations. However, Armstrong's objections, if valid, equally affect all mentalist construals of common ground. Armstrong argues, first, that many creatures that probably lack meta-representational capacities (like baboons and human infants) nevertheless engage in interactions that would appear to require common ground in some substantive sense. Secondly, he argues that human adults, who surely have meta-representational capacities, do not routinely employ them in their interactions because meta-representation is cognitively demanding. If this is correct, the common ground between any two individuals is intermittent, at best, which doesn't seem right.

Armstrong takes these arguments to imply that the mentalist view on common ground draws too sharp a line between human adults and other creatures (including human infants and baboons), and therefore stands in the way of a properly gradualist theory of the evolution of social and communicative interactions. A more inclusive account is called for.

While Armstrong's own approach "retains the psychological orientation" of the meta-representational view,

> [. . .] it focuses on a much more basic set of psychological states. The animating idea of the account is that common ground consists in a set of mutual attitudes that a group of agents accept because of a history of reciprocal cognitive responsiveness to the presence of one another and to the states of the world more generally. (Armstrong, 2025: 124–125)

In the jargon of this chapter, Armstrong uses the adjective "mutual" to mean "shared", not "mutual"; his "mutual attitudes" are instances of first-order sharing, hence not instances of common ground as we have defined that term. For our purposes, the main features of Armstrong's approach are the following. First, it is resolutely mentalist in the sense that common ground is defined in such psychological terms as "mental representation" and "cognitive responsiveness"; normative varieties of common ground aren't invited. In this respect, Armstrong's notion of common ground is narrower than mine. Secondly, to the extent that Armstrong's notion is broader than its mentalist kin, its scope is broader than mine. However, thirdly, it is still limited to creatures with relatively advanced psychological resources, "which display the underlying capacities for social representation and social memory", and enable their owners to represent others as "individually distinctive agents" (p. 125). Baboons and human infants can have Armstrong-style common ground; bacteria and butterflies cannot.

Armstrong's fundamental insight is that all forms of first-order sharing aren't alike, because there are some that enable social interactions. Following this lead, I argue that there is room and need for a notion that is more inclusive than Armstrong's common ground, as well as mine. I call it "shared ground". The overall picture is this:

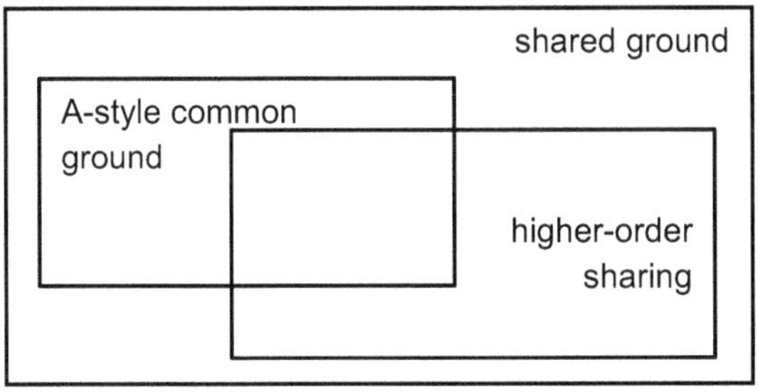

Armstrong's notion of common ground and mine cover intersecting areas of shared ground. Shared ground is broader than both in that it accommodates comparatively primitive creatures like bacteria and butterflies. My notion of common ground is broader than Armstrong's in that it subsumes normative forms of higher-order sharing, and narrower in that it doesn't cater for any kind of first-order sharing. In the following, the terms "common ground" and "higher-order sharing" are used as synonyms by default unless Armstrong's notion of common ground is at issue.

8.4 Shared Ground

It is agreed on all hands that common ground helps social creatures to coordinate their actions. Pragmatics is a special case: interlocutors rely on common ground to coordinate their utterances. But what does it mean to coordinate one's actions in general? Game theory offers an answer (O'Connor, 2020; Mühlenbernd & Baumann, 2025). It invites us to view action coordination as a special kind of interdependent decision making. For example, Barney and Betty enter a room together. Who goes first? The couple each has two strategies to choose from: go first or go second. If both go first, they will get stuck, because the door isn't wide enough for two people to enter at the same time. If both decide to go second, each will be waiting for the other. Complementary strategies are called for: either Barney goes first and Betty second, or the other way round. Either combination of strategies is acceptable to both, but each must decide on a strategy without knowing in advance what the other will do. Barney and Betty have a coordination problem, and if they manage to solve it, they will have coordinated their actions.

Coordination problems are ubiquitous in human society and throughout nature. They occur in a rich variety of forms, countless solutions for which have been invented and evolved. Usually, these solutions involve *correlation devices:* circumstances external to the interaction as such that participants conditionalize their actions on, knowingly or unknowingly (Vanderschraaf, 1995; Brandenburger & Friedenberg, 2008; Guala, 2016: Chapter 4). Suppose Barney and Betty are an old-fashioned couple for whom the rule "Ladies first" is normative. Then their actions will tend to conform to that rule: Betty goes first and Barney second. They conditionalize their strategies on the rule, using it as a correlation device. The rule is also common ground between Barney and Betty, and this is clearly not an accident: common ground is a correlation device, or rather, a conglomerate of correlation devices. But as we will shortly see, there are many correlation devices that don't involve higher-order sharing, and therefore are not common ground, but nonetheless facilitate coordinated action. Thus, the notion of a correlation device affords a general perspective on action coordination, in which common ground is a special case.

Let's say that something is "shared ground" for a group iff it serves as a correlation device for that group. So the predicate of being shared ground is coextensive with the predicate of serving as a correlation device. I am introducing two coextensive terms here because they will help to distinguish between two forms of higher-orderness. First, higher-order *sharing*, that is, common ground, is a special kind of shared ground, which is among the protagonists of my story. Second, a higher-order *correlation device* is a special kind of correlation device, which makes cameo appearances in Sections 8.8 and 8.11. These two predicates are not coextensive.

As defined here, shared ground covers all varieties of common ground discussed in the foregoing, including Armstrong's, but in addition, it covers forms of first-order sharing that fall outside Armstrong's common ground. In this perspective, common ground is a recent development in the long evolution of coordinated action. Over many millions of years, correlation devices gradually became more complex, enabling increasingly powerful forms of coordinated action, until the first forms of higher-order sharing arrived on the scene, and common ground was born. In the following, we will sample this immense sprawl of developments by discussing a handful of instances of action coordination ranging in sophistication from low-end (bacteria and butterflies) to high-end (human adults). The purpose of this discussion is not only to exemplify the sheer variety of circumstances that can be shared ground, but also to show that the concept of shared ground reveals continuities underlying otherwise disparate forms of coordinated action.

8.5 Quorum Sensing

The Hawaiian bobtail squid (*Euprymna scolopes*) lives in symbiosis with luminescent bacteria (*Aliivibrio fischeri*) lodged in a light organ on its underside. The squid provides its symbionts with nutrients, in return for which the bacteria emit light that reduces the squid's silhouette when viewed from below, thus helping it to camouflage against predators. As if that wasn't impressive enough, the bacteria light up in unison and only when they are sufficiently numerous (Verma & Miyashiro, 2013). They achieve this feat by a process called "quorum sensing". The bacteria continuously secrete and have receptors for "signalling molecules". As long as the bacterial density is low, the density of signaling molecules will be too low to be registered, but when the bacteria are sufficiently numerous for the molecules to be detected, all bacteria respond by emitting light and secreting more signaling molecules. As a result, they all light up together.

Aliivibrio fischeri coordinate their behaviors with the help of the density levels of signaling molecules. These levels serve as correlation devices and thus are shared ground. They are not common ground because, in the bacterial world, all sharing is of the first order. Lacking minds as well as normative practices, bacteria don't engage in higher-order sharing of any kind. Density levels are not common ground in Armstrong's sense, either, if only because bacteria don't represent each other as individual agents. Hence, this case illustrates how basic and therefore how general the notion of shared ground really is.

Given that, in the habitat created by the bobtail squid, the density of signaling molecules strongly correlates with the density of *Aliivibrio fischeri*, both the former and the latter may be viewed as correlation devices, with molecular density serving as a proximal and bacterial density as a distal correlation device. Although from a functional point of view, molecular density is just a proxy for bacterial density, the two measures are practically equivalent, and therefore, in this case, the proximal/distal distinction has little work to do. By contrast, we will see that, in the case of common ground, the distinction is crucial (Section 8.9).

Quorum sensing is often represented as a form of communication, as witness the term "signalling molecule" and Bassler's (1999) title: "How bacteria talk to each other". However, the so-called "signalling molecules" secreted by individual bacteria are not signals. Rather, it is the *density* of signaling molecules that serves as a correlation device. Individual secretions impact density levels, but are not themselves correlation devices.

8.6 Territoriality

Territoriality is a trait of those species in which individuals or groups may claim an area and defend it against intruders. Territorial behaviors serve to monopolize food sources, protect nesting sites, or attract mates. If an animal enters the territory of a conspecific, be it on purpose or by accident, both face the same dilemma: stay or leave. If both parties stay, there is likely to be a fight, and both may be injured. If both leave, they equally lose, though perhaps not as badly as when they fight. Therefore, on balance, the best outcome is that one leaves while the other stays. Surprisingly, fighting ability turns out to be a poor predictor of how territorial conflicts are decided, for the resident wins in the majority of cases. Usually, it is the intruder that withdraws after a more or less perfunctory skirmish. Thus, residential status serves as a correlation device.

Although "territoriality" may connote long-term occupation and patrolling of sizeable plots of real estate, territories can be short-lived and small. My favorite example is the speckled wood butterfly, as described by Davies (1978). These butterflies spend the night in the tree tops, but when the sun starts casting spots of light on the forest floor, the males descend to occupy a sunspot that provides the warmth that they need to remain active during the day. Whenever another male enters an occupied sunspot, resident and intruder engage in a spiral upward flight, after which one of them returns to perch in the sunspot, while the other flies away. Initially, Davies interpreted these spiral flights as contests until he found, after many observations, that it was always the intruder that withdrew:

> This applied even when the owner was a very tatty male whose wings were torn and the intruder was a perfect male in mint condition. Thus the spiral flights, only a few seconds in duration, cannot be regarded as contests at all. Rather they are a short conventional display where, to put it anthropomorphically, the owner says "I was first here" and the intruder says "Sorry, I didn't know there was anyone occupying the sunspot, I'll retreat back to the canopy". (Davies, 1978: 145)

In the butterfly population studied by Davies, sunspots were allocated on a first-come, first-served basis, and Davies managed to show that "ownership" expired within seconds after a sunspot was vacated. Hence, what territoriality comes down to in this case is that, when two males find themselves in the same sunspot, their behavioral strategies are conditional on which of them arrived first. That simple fact is shared ground between them.

I take it that butterfly sociality is not premised on either normativity or mental-state attribution, so there can be no higher-order sharing between butterflies. Nor do they meet the psychological requirements for Armstrong-style common ground. Still, we see pairs of butterflies using the same fact to solve a coordination problem between them. They have shared ground but not common ground in any of the modes considered in the foregoing.

Of course, territoriality is not confined to butterflies. We have our territorial moments, too. Picture yourself in a cinema, waiting for the feature film to begin. Seating is free, but then this impudent woman shows up and claims "your" seat. How will you react? Speaking for myself, I would pointedly refuse and appeal to the same rule that governs the competition for sunspots in butterflies: "First arrivals take precedence". Even if, unlike the butterflies, I would be articulate and normative about it, the principle is the same; apparently, it has deep evolutionary roots. In the upcoming section, I argue that it is a convention in humans and other species, and that conventions are shared ground in all cases, including butterflies, though in adult humans, they are common ground, as well.

8.7 Precedents and Conventions

We are creatures of habits as well as rules, and probably more of the former than we would care to admit. All too often, we act in such-and-such a way because we did so in the past. Imagine, for example, a small workshop with a dozen or so participants convening on two consecutive days. On Thursday, the allocation of seats emerges semi-randomly from the participants' idiosyncratic choices and the order in which they arrive. By contrast, on Friday, it is as if a seating arrangement had been agreed upon in advance: everyone is sitting where they sat the day before. Thus, a recurring coordination problem is solved on the strength of a single precedent. Precedents are correlation devices, hence shared ground; in this case, the precedent is common ground as well.

The power of precedents is even stronger in the following thought experiment (Geurts, 2018). We invite two people who haven't met before, seat them in separate rooms, and give them identical instructions:

> We ask you and the person next door to choose an animal species. If both of you choose the same species, you earn five euros each; otherwise you get nothing.

This is a "pure" coordination game, in which there is no conflict of interest: coordination is the sole objective, and neither player prefers any solution over another. Suppose that our subjects strike lucky: both choose llamas and receive five euros each. Then we repeat the experiment with the same subjects. Which species will they choose now? Naturally, both will choose llamas again and will expect the other to make the same choice. What is more, they will agree that it would have been "wrong" to choose differently.

On the first trial, the coordination problem was solved by luck. On the second trial, it was solved on the strength of a single precedent, and with normative expectations about the outcome. However, in general, it takes more than one precedent to solve a coordination problem. If a coordination problem keeps recurring and the same solution is achieved on sufficiently many occasions, a behavioral pattern emerges. A convention is born. Conventions, thus understood, are precedent types, which many species employ as correlation devices.

This notion of convention is vague because it is impossible to say, in general, how many precedents it takes to make a convention. More importantly, it is a minimal notion of convention in the style of Skyrms (1996) and Millikan (1998). Lewis (1969) famously defined conventions as behavioral patterns that are common ground in a population, in a normative sense of "common ground".[1] Lewisian conventions are a lot more demanding of psychological resources than the minimal ones canvassed here, and have been lambasted for that reason (Skyrms, 1996; Millikan, 1998; Armstrong, 2025). However, theoretical concepts should not be discarded merely because they apply to humans alone, and it seems entirely plausible to me that Lewis's concept of convention, or at least something in its spirit, captures an important aspect of human sociality (Geurts, 2018, 2024; van Tiel & Geurts, 2025).

At the end of the last section, I claimed that, while "First arrivals take precedence" is a convention in many species, it is special for humans. That claim can now be made more precise: "First arrivals take precedence" is a minimal convention in many species and a Lewisian convention in ours. Put otherwise, it is shared ground in general, and common ground for us. What began as a regularity on which many species conditionalize their behavioral strategies, became a normative rule that people rely on in their interactions. This is one instance of a more general pattern. It is well established that, from an early age, humans are prone to be normative about regularities in their interactions (Schmidt & Rakoczy, 2019; Przepiorka et al., 2022). We tend to upgrade such regularities to social norms. This is one of the pathways in the evolution from shared ground to common ground.

8.8 Signaling and Language

Butterflies, baboons, and bankers: they all use signals to communicate with conspecifics, and occasionally with heterospecifics, too. Signaling is rife throughout the animal kingdom, and formal models confirm that the conditions for signaling to evolve are not particularly stringent (Skyrms, 1996, 2010; O'Connor, 2020; Mühlenbernd & Baumann, 2025). Signals come in many forms (vocal, gestural, facial, olfactory, etc.) and are used for all sorts of purposes: to announce good food sources, warn against predators, attract mates, and so on. Intuitively, it seems plausible that signals are correlation devices, and with the help of Lewis's (1969) signaling games, this intuition can be made more precise.

Signaling games are coordination games with players that either send or receive signals. In the standard set-up, senders observe their surroundings and produce signals so as to enable receivers to act adequately, given the observed state of affairs. For a simple example, consider a scenario with one sender that observes either a leopard or an eagle; one receiver that is best advised to climb up a tree, in the first case, or hide under a bush, in the second; and two signals available to be used as alarm calls: barking and whistling. Assuming that it is in the interest of both parties that the receiver act appropriately, it is clear that the sender and the receiver should settle for one of the following two combinations of strategies:

	Sender		**Receiver**	
	If I see . . .	then I . . .	If I hear . . .	then I . . .
(1)	a leopard	bark	a bark	climb up a tree
	an eagle	whistle	a whistle	hide under a bush
(2)	a leopard	whistle	a whistle	climb up a tree
	an eagle	bark	a bark	hide under a bush

In Lewis's terminology, these two combinations are "signalling systems": senders and receivers will do equally well by adopting either of these, and worse by adopting any other combination of strategies. Even in this simple scenario, there are 16 possible combinations of strategies, only two of which are signaling systems, and the size of the problem increases exponentially with the number of stimulus situations, signals, and responses. Nevertheless, signaling systems have evolved or been invented many times and throughout nature.

A signaling game presents agents with a coordination problem that they can solve by making their behavioral strategies conditional on a signaling convention. Usually, such conventions emerge from past interactions, and the evolutionary process may involve Darwinian selection, social learning, or both. Humans are special in that they are capable of establishing conventions by means of deliberation. In any case, signaling conventions are correlation devices, and therefore shared ground between sender and receiver.

Although it has been a long way from animal signaling to language-based communication, the latter evolved out of the former and retains some of its core features. Basically, languages are conventional systems for constructing compound signals, like "does your dog bite". In other words, they are higher-order correlation devices: a language enables its users to agree on correlation devices for them to coordinate their communicative exchanges. Hence, the language we speak with one another is shared ground between us. But although it is often held that a shared language is common ground, too, that is only partly true, as the following case illustrates. It is common ground between speakers of English that *s* is pronounced differently in "dogs" and "boys" than it is in "cats" and "girls". As it turns out, there is a regular pattern here:

- When preceded by a voiced sound, *s* is pronounced \z\.
- When preceded by an unvoiced sound, *s* is pronounced \s\.

These are conventional rules of English, which don't hold even in closely related languages like Dutch or German, for instance. They are routinely applied to new words, but only implicitly. More generally, while the rules of a shared language are shared ground, only some of their instances will be common ground.

While languages are higher-order correlation devices for constructing compound signals, words are higher-order correlation devices, too, though in a different and more basic sense. What is informally called "the meaning" of a word is in fact a cluster of precedent types, that is, ways in which the word is commonly used. For example, the English word "box" is commonly used to refer to a container with a flat base and sides, a rectangular field on a page or a computer screen, the penalty area in football, an enclosed space in a theatre reserved for some well-heeled party, etc. In an actual use of the word, one of these "senses" is selected to assist in coordinating the ongoing interaction. In this respect, some non-human signals resemble words. Many of the communicative gestures used by chimpanzees have multiple functions (Graham & Hobaiter, 2025) and, like words, gesture types serve as higher-order correlation devices that help their users to settle on a single function for a given token.

8.9 From Shared to Common Ground

Shared ground is a highly inclusive concept that encompasses a great variety of instances. Thus far, our focus has been chiefly on life forms whose limited social practices and cognitive capacities fail to support higher-order sharing of any kind, though at some points we have already glimpsed at the other end of the zoological spectrum, where we find creatures that do have the required capacities and practices. Adult humans are the paradigm, and they are the focal subjects of the remaining sections. They rely on various modes of common ground, which apparently boost their coordinative capacities. But how?

Let me try to answer that question by means of a few simple thought experiments, starting with a replication of the coordination game of Section 8.7. Two subjects sitting in different rooms are asked to choose an animal species; if they chance upon the same species, they win; otherwise, they lose. Then another pair of subjects is given the same task, only they are sitting at a table facing each other, with a plastic effigy of an elk between them. Surely, the second pair will be significantly more successful than the first; the plastic elk enables them to converge on the same response with great confidence. Both subjects are certain that "elk" is the response that uniquely stands out, and both are certain that they share that certainty. Hence, it is common ground between them that "elk" is the right response.

In the second experiment, the plastic elk serves as a fail-safe correlation device, which was lacking in the first. But it is also true that our subjects' response strategies are contingent on their common ground. Apparently, there are two correlation devices involved: one distal (the plastic elk on the table) and one proximal (the common ground), and the two are obviously related: the plastic elk makes the species salient, and not just to both subjects separately, for they have higher-order certainty. Both the salient stimulus and their common ground are needed to explain how our subjects managed to converge on the same response.

Thus far, the case is analogous to that of the Hawaiian bobtail squid and its luminescent symbionts, in which bacterial and molecular density serve as distal and proximal correlation devices, respectively (Section 8.5). However, the present case is importantly different, because now there is no strong correlation between proximal and distal correlation devices, for the simple reason that the plastic elk is dispensable; it is not needed to make it common ground that "elk" is the right response. If no species in particular had been salient at the outset, subjects could still have converged on the elk, for example, by talking to each other:

S1: "How about the elk?"
S2: "Okay".

More generally, there are indefinitely many ways in which it may become common ground that φ, for any φ. In the case of the bobtail, molecular density is a proxy for bacterial density. By contrast, in the second version of our thought experiment, common ground is not a proxy for the plastic elk, even if it is common ground that it is on the table. Indeed, in this case, common ground doesn't seem to be a proxy for anything.

It may be tempting to think that this contrast is due to the mentalist mode of common ground we adopted by supposing that there was mutual *certainty* that there was a plastic elk on the table. Like "believe", "be certain" is a non-factive predicate: one can be certain that φ regardless of whether φ is true or not. However, non-factivity is by no means unique to psychological attributes. For example, no matter whether "commitment" is construed in social or in psychological terms, one's commitment to the truth of "The earth is flat" does not entail that the earth is flat. More generally, it is an essential feature of common ground that it will serve as a correlation device even if its implications are questionable or worse. There may be a consensus that the Pope is infallible, that a bridge is as strong as it looks, or that tariffs are beneficial, for example, and in each case, the common ground may enable action coordination despite the fact that the consensus is misplaced.

For coordination to succeed, mutual agreement may be more urgent than being right, and that is what non-factivity allows for. This is the first core feature of common ground. The second is its staying power: once something is common ground, it may remain so for a longer period of time. Common ground is able to persist, and persistence is essential for *prospective* coordination, which was one of the major developments in our lineage.

8.10 Commitments and Prospective Coordination

Our lineage has taken action coordination to another level by adopting a thoroughly normative lifestyle. We treat each other as having deontic statuses: as being bound, allowed, or forbidden to act in certain ways. Since, by and large, we live up to our deontic statuses, this lifestyle revolutionized action coordination (Geurts, 2019, 2022). With the advent of deontic statuses, a new mode of common ground arrived on the social scene, providing us with correlation devices of unprecedented power.

When I enter a bakery, deontic statuses are readjusted. I am now a customer, committed to behaving as a customer should, and entitled to be treated as one by the shop assistant as well as the other customers. I have to wait my turn, the others are entitled to be served before me, and when it becomes my turn to be served, I am committed to select between the products on offer, the shop assistant is committed to hand over the items I wish to buy, and so on. All this and more is now part of the deontic statuses of those present, and it is common ground between us.

For all its mundanity, the bakery scenario exemplifies how normativity pervades our social interactions. This holds for our communicative interactions, too, and with a vengeance. While language is used in many ways, its primary function is to negotiate deontic statuses, and in particular to undertake and share commitments. This is what Brandom (1994) calls "normative pragmatics".[2] If I promise to read your paper, I become committed to read your paper; if I ask you to read my paper, my objective is that you become committed to read my paper; if we make a lunch appointment, we both become committed to meet at a given time and place and have lunch together; if I tell you that Lenin was a vegetarian, I commit myself to the truth of the proposition that Lenin was a vegetarian; and so on.

Deontic statuses are constantly adjusted in the wake of our interactions, communicative and non-communicative. But at the same time, they are basically stable. Suppose Betty and Wilma agree to see the remastered edition of "Rocky I" together. As a result of their communicative exchange, both women become committed to meet at the local cinema at a given time, and until then their newly instituted commitments persist by default. They are not written in stone, but unless either party moves to renegotiate, their commitments are with them to stay. It is this combination of negotiability and persistence that makes deontic statuses such powerful correlation devices, which empower us to flexibly coordinate our activities across timespans of practically indefinite length.

How does common ground enter into this picture? In earlier work, I have developed a theory that ties commitment and common ground together in an exceptionally strong way (Geurts, 2019, 2024). Here I briskly summarize the main ideas. In Section 8.2, mutual commitment was defined as follows: A and B are mutually committed to the truth of φ iff:

(1) A and B are committed to the truth of φ.
(2) A and B are committed to the truth of (1).
(3) A and B are committed to the truth of (2), and so on.

Although this definition treats commitments as attributes, properly understood, they are relational. If I promise to read your paper, I become committed *to you* to read your paper. Hence, commitments have not two but three parameters: two individuals and a propositional content. Since this makes them a bit unwieldy to handle, let's use $C_{A,B}\varphi$ as a shorthand for representing that A is committed to B to the truth of φ. Recast in this notation, A and B are mutually committed to the truth of φ iff:

(1) $C_{A,B}\varphi$ and $C_{B,A}\varphi$.
(2) $C_{A,B}(1)$ and $C_{B,A}(1)$.
(3) $C_{A,B}(2)$ and $C_{B,A}(2)$, and so on.

Now we introduce the following constraint on commitments: $C_{A,B}\varphi$ doesn't hold unless B accepts that A is thus committed. More formally, and construing acceptance in terms of commitment, the proposed constraint reads:

$C_{A,B}\varphi$ only if $C_{B,A}C_{A,B}\varphi$ (Acceptance principle).

Given the Acceptance principle, it follows immediately that (1) entails (2), (3), etc., and therefore we can cut back our definition of mutual commitment to:

A and B are mutually committed to the truth of φ iff $C_{A,B}\varphi$ and $C_{B,A}\varphi$.

If we read "$C_{A,B}\varphi$ and $C_{B,A}\varphi$" as meaning that A and B share a commitment to the truth of φ, commitment sharing and mutual commitment are equivalent. Put otherwise, on the proposed account, the distinction between first-order and higher-order sharing collapses; commitment sharing *is* a mode of common ground.

If this result seems puzzling at first, note that, while it is caused by the Acceptance principle, it is equally important that commitments are *relations* between individuals and that commitment sharing is *symmetric*. It is only in conjunction with these features that the Acceptance principle levels the distinction between first-order and higher-order sharing. To highlight the importance of symmetry, let's compare commitment sharing with belief sharing. Suppose that both Betty and Wilma believe that Stalin was gay and that neither is aware that they share that belief; so this is an instance of strictly first-order sharing. Meanwhile, Fred and Barney share a commitment to the truth of "Stalin was gay". Theirs is first-order sharing, too, but it is symmetric in the sense that each is committed to the other. So even before the Acceptance principle comes into play, commitment sharing is markedly different from belief sharing, and that is not only because commitments are normative while beliefs are not.

At the end of the last section, I noted that, due to its ability to persist, common ground underpins prospective coordination. In this section, I argued that our proclivity for attributing deontic statuses to one another brought about a mode of common ground that is not only persistent but flexible as well: we use language to negotiate commitments on the fly. When Betty and Wilma agreed to go to the cinema together, they communicated at time t_0 to negotiate a shared commitment that caused them to participate in a joint activity at some future time t_1, and that shared commitment was common ground between them all the while from t_0 to t_1. Thus, inter-human communication, deontic statuses, and common ground come together in our account of prospective coordination, which enabled a rampant variety of social entities that are essentially persistent, from marriages and money to the Holy Church of Rome.

8.11 Norms and Institutions

Much of the discussion about normativity has been about norms and has failed to separate them from deontic statuses. That is unfortunate, because the distinction is essential. Norms are usually seen as rules or regularities that are comparatively static and generic in the sense that they govern interactions within larger groups. Deontic statuses are neither static nor generic. If Fred promises Wilma to do the dishes, their deontic statuses change: he is now committed to doing the dishes, and she is entitled to act on the premise that he will do so. His commitment and her entitlement are short-lived, non-generic relations between the two of them. By contrast, the rule "Keep your promises" is long-lived and generic, and if it "governs"

Fred's and Wilma's interactions at all, it does so indirectly, by enabling them to negotiate their deontic statuses (Geurts, 2025).

Once the distinction has been made, it isn't hard to see that deontic statuses and norms don't always align. For example, the rule that promises ought to be kept is enforced, if at all, with varying degrees of rigor, depending on the people involved, their social stations and relationships, and other factors. Some norms are observed more strictly than others, but all norms are violated at least some of the time, and in many cases, we are *entitled* to violate a norm. This also holds for allegedly absolute norms such as "Don't kill people" or "Don't eat people". In brief, it depends on the context what, if any, commitments a norm prompts us to undertake. Norms are second-order correlation devices on which we can, and often will, conditionalize our strategies for adopting deontic statuses, which are first-order correlation devices.

Norms tend to hang together in larger social structures: clubs, conferences, hospitals, marriages, orchestras, schools, and supermarkets (Guala, 2016). The list is long and varied, but what all institutions have in common is that each has its own norms for coordinating the actions of its participants. If norms are second-order correlation devices, institutions are third-order correlation devices. And that is not the end of the progression, as there are norms and institutions that regulate norms and institutions. For example, many countries have educational laws issued by the government to regulate schools and their regulations, which govern the deontic statuses of pupils and teachers.

As if all this layering of coordination and normativity wasn't enough, it gets even better. As discussed by Hindriks & Guala (2021), institutions are hybrid entities that not only create norms but are also governed by them, have objectives and values, organizational structures that may be exceedingly complex, and so on. But the most striking aspect of all is that many of them are agreed to be *agents*. We treat universities, for example, as having goals that guide their actions (Bratman, 2022: Stoelhorst, this volume); we interact with them as professors, students, alumni, etc., and universities interact with each other and with other agentive institutions, like municipalities, newspapers, and ministries. Finally, and perhaps most perplexingly, they have deontic statuses: agentive institutions are treated as oversized persons with commitments and entitlements.

Thus, what began as a momentous though, at the time, inconspicuous innovation in coordinating small-scale interactions culminated in the personification of large, multi-dimensional, higher-order correlation devices, which not only *are* common ground between us, but also *have* common ground *with* us and between one another. Such are the powers of normativity.

8.12 Conclusion

We have come to the end of our scenic tour of correlation devices. Our starting point was threefold. First, there is a broad consensus that common ground, understood as higher-order sharing, is foundational to human sociality. Second, the primary purpose of common ground is action coordination. Third, as argued by Armstrong (2025), even the humblest forms of common ground may be too demanding for non-human organisms to achieve. Taken together, these premises prompted us to ask what enables action coordination in the absence of common ground and what, if anything, is shared between coordinated activities with and without common ground.

The proposed answer was, in a word, "correlation devices". At one end of the coordination spectrum, bioluminescent bacteria synchronize their behaviors by using their colony's density as a correlation device. This is a primitive form of sharing that enables a group to coordinate their activities without taking heed of each other at all. At the other end, humans coordinate their activities by taking into account each other's beliefs, commitments, and so on, using common ground as a correlation device.

While shared ground is primordial, the first modes of common ground arrived late by evolutionary standards. But common ground was a game changer, which enabled Lewisian conventions (Section 8.7), language (Section 8.8), a distinctively normative style of sociality and communication (Section 8.10), and norms and institutions (Section 8.11). Human societies with their rich social ontologies wouldn't have existed without the advanced forms of action coordination afforded by common ground, among which prospective coordination stands out. When our ancestors began treating each other as having deontic statuses that could be negotiated by communicative means, it became possible to coordinate activities across time spans of ever-increasing length. The glorious and inglorious products of prospective coordination are all around us.

Acknowledgements

At several stages during its long gestation, this chapter profited a great deal from comments and questions by Josh Armstrong, Filip Buekens, Mathijs Geurts, Wil Martens, Thomas A.C. Reydon, Antonio Scarafone, Ric Sims, Marc Slors, J.W. Stoelhorst, and Martina Valković. Thank you all.

Notes

1 Actually, the mode of common ground Lewis employed was "mutual reasons to believe", with infinitely many orders of reasons to believe (Section 8.2).
2 Besides Brandom, the normative-pragmatics family also includes Walton & Krabbe (1995), Carassa & Colombetti (2009), Krifka (2022), Geurts (2019, 2022), and many others.

References

Armstrong, J. (2018): Provincialism in pragmatics, *Philosophical Perspectives* 32: 5–40.

Armstrong, J. (2025): The evolutionary foundations of common ground, in: Geurts, B. & Moore, R. (Eds): *Evolutionary Pragmatics*, Oxford: Oxford University Press, pp. 111–138.

Bach, K. & Harnish, R.M. (1979): *Linguistic Communication and Speech Acts*, Cambridge, MA: MIT Press.

Bassler, B.L. (1999): How bacteria talk to each other: Regulation of gene expression by quorum sensing, *Current Opinion in Microbiology* 2: 582–587.

Brandenburger, A. & Friedenberg, A. (2008): Intrinsic correlation in games, *Journal of Economic Theory* 141: 28–67.

Brandom, R. (1994): *Making It Explicit*, Cambridge, MA: Harvard University Press.

Bratman, M.E. (2022): *Shared and Institutional Agency: Toward a Planning Theory of Human Practical Organization*, New York: Oxford University Press.

Carassa, A. & Colombetti, M. (2009): Joint meaning, *Journal of Pragmatics* 41: 1837–1854.

Clark, H.H. (1996): *Using Language*, Cambridge: Cambridge University Press.

Clark, H.H. (2006): Social actions, social commitments, in: Enfield, N. & Levinson, S.C. (Eds): *Roots of Human Sociality: Culture, Cognition and Interaction*, Oxford: Berg, pp. 126–150.

Clark, H.H. & Marshall, C. (1981): Definite reference and mutual knowledge, in: Joshi, A., Webber, B.L. & Sag, I. (Eds): *Elements of Discourse Understanding*, Cambridge: Cambridge University Press, pp. 10–63.

Davies, N. (1978): Territorial defence in the speckled wood butterfly (*Pararge aegeria*): The resident always wins, *Animal Behaviour* 26: 138–147.

Geurts, B. (2018): Convention and common ground, *Mind and Language* 33: 115–129.

Geurts, B. (2019): Communication as commitment sharing: Speech acts, implicatures, common ground, *Theoretical Linguistics* 45: 1–30.

Geurts, B. (2022): Evolutionary pragmatics: From chimp-style communication to human discourse, *Journal of Pragmatics* 200: 24–34.

Geurts, B. (2024): Common ground in pragmatics, in: Zalta, E.N. & Nodelman, U. (Eds): *The Stanford Encyclopedia of Philosophy* (Winter 2024 Edition), https://plato.stanford.edu/archives/win2024/entries/common-ground-pragmatics/.

Geurts, B. (2025): Normative pragmatics and social structures: An evolutionary perspective, in: Geurts, B. & Moore, R. (Eds): *Evolutionary Pragmatics*, Oxford: Oxford University Press, pp. 226–245.

Gilbert, M. (2014): *Joint Commitment: How We Make the Social World*, New York: Oxford University Press.

Graham, K.E. & Hobaiter, C. (2025): Pragmatics in ape gesture, in: Geurts, B. & Moore, R. (Eds): *Evolutionary Pragmatics*, Oxford: Oxford University Press, pp. 139–155.

Guala, F. (2016): *Understanding Institutions: The Science and Philosophy of Living Together*, Princeton: Princeton University Press.

Hindriks, F. & Guala, F. (2021): The functions of institutions: Etiology and teleology, *Synthese* 198: 2027–2043.

Krifka, M. (2022): Adjacency pairs in common ground update: Assertions, questions, greetings, offers, commands, in: Gregoromichelaki, E., Hough, J. & Kelleher, J.D. (Eds): *Proceedings of the 26th Workshop on the Semantics and Pragmatics of Dialogue*, Dublin, SEMDIAL, pp. 94–102.

Lewis, D.K. (1969): *Convention*, Cambridge, MA: Harvard University Press.
Millikan, R.G. (1998): Language conventions made simple, *Journal of Philosophy* 95: 161–180.
Mühlenbernd, R. & Baumann, A. (2025): Population-level models of evolutionary pragmatics, in: Geurts, B. & Moore, R. (Eds): *Evolutionary Pragmatics*, Oxford: Oxford University Press, pp. 196–225.
O'Connor, C. (2020): *Games in the Philosophy of Biology*, Cambridge: Cambridge University Press.
Przepiorka, W., Szekely, A., Andrighetto, G., Diekmann, A. & Tummolini, L. (2022): How norms emerge from conventions (and change), *Socius* 8: 1–16.
Schiffer, S. (1972): *Meaning*, Oxford: Oxford University Press.
Schmidt, M.F.H. & Rakoczy, H. (2019): On the uniqueness of human normative attitudes, in: Roughley, N. & Bayertz, K. (Eds): *The Normative Animal? On the Anthropological Significance of Social, Moral, and Linguistic Norms*, Oxford: Oxford University Press, pp. 122–135.
Skyrms, B. (1996): *Evolution of the Social Contract*, Cambridge: Cambridge University Press.
Skyrms, B. (2010): *Signals: Evolution, Learning, and Information*, New York: Oxford University Press.
Stalnaker, R. (1978): Assertion, in: Cole, P. (Ed): *Radical Pragmatics*, New York: Academic Press.
Stalnaker, R. (2014): *Context*, New York: Oxford University Press.
van Tiel, B. & Geurts, B. (2025): Conventions, coordination, and arbitrariness, in: Geurts, B. & Moore, R. (Eds): *Evolutionary Pragmatics*, Oxford: Oxford University Press, pp. 179–195.
Vanderschraaf, P. (1995): Convention as correlated equilibrium, *Erkenntnis* 42: 65–87.
Vanderschraaf, P. & Sillari, G. (2023): Common knowledge, in: Zalta, E.N. & Nodelman, U. (Eds): *The Stanford Encyclopedia of Philosophy* (Winter 2023 Edition), https://plato.stanford.edu/archives/win2023/entries/common-knowledge/.
Verma, S.C. & Miyashiro, T. (2013): Quorum sensing in the squid-vibrio symbiosis, *International Journal of Molecular Sciences* 14: 16386–16401.
Walton, D.N. & Krabbe, E.C. (1995): *Commitment in Dialogue: Basic Concepts of Interpersonal Reasoning*, Albany: State University of New York Press.
Yalcin, S. (2024): Defining common ground, *Linguistics and Philosophy* 47: 1045–1070.

9 Collective Intentionality

From Social Ontology to Cultural Evolution and Back

Marc Slors

9.1 Introduction

There are not many instances where the debate on cultural evolution[1] and the debate on social ontology directly influence each other. One widely published exception, though, is the concept of collective intentionality that is taken from social ontology and applied in the domain of cultural evolution by Michael Tomasello and others (Angus & Newton, 2015; Boehm, 2018; Herrmann et al., 2007; Rakoczy, 2009; Tomasello, 2014, 2019a, 2019b). This case of cross-fertilization between debates is presented in a one-directional way, as social ontology influencing research on cultural evolution (Tomasello, 2016). And that is the usual frame in which this overlap between areas of research is assessed; the main question being whether Tomasello's use of the concept is still in line with its philosophical origins, and if so, which theory of collective intentionality it is most congenial with (e.g., Angelino, 2022; Pettit, 2020). In this chapter, I want to look at the possibility of a reverse direction of influence: what lessons, if any, can philosophers learn from views on the evolution of collective intentionality, understood in Tomasello's terms, as well as those who hold congenial views?

I will argue that Tomasello's application of the notion of collective intentionality in the context of cultural evolution offers something that is, so far, absent in the social ontology literature: a genuine, non-mysterious form of "top-down" influence of group-level processes on the individuals that collectively make up institutions and societies – the entities to which collective intentionality is attributed. Collective intentionality, as described by Tomasello and colleagues, implicitly involves suprapersonal, transgenerational, self-organizing cycles of socialization that set individuals up as functioning group members who, in turn, will socialize their offspring in a similar fashion. These cycles can, as I will argue, best be explained in terms of cultural group selection.

Social ontologists are highly suspicious of self-organizing suprapersonal processes. They acknowledge that collective intentionality is more than an

DOI: 10.4324/9781003643845-12

aggregate of individual intentions. But they also take care to avoid group-to-individual, top-down determination on the basis that this implies either a mysterious "hive-mind" or objectionable historical teleology. I will argue that cultural group selection of self-organizing cycles of socialization is metaphysically bona fide; it has nothing to do with hive-minds or historical teleology. But it does deliver genuine group-to-individual top-down influence. Whether this means that Tomasello's notion of collective intentionality has something to offer that may also be of use in the domain of social ontology is a question that I will discuss at the end of this chapter. I will argue for a cautiously optimistic answer, and hence for a bidirectional influence between the debates on cultural evolution and social ontology.

The chapter is set up as follows. In the next section, I will very briefly introduce the notion of collective intentionality. In Section 9.3, I will discuss Tomasello's adoption of the concept and the way in which he has put it to use in the context of cultural evolution. In Section 9.4, I will discuss Tomasello's notion of "group-mindedness" as an integral part of his account of collective intentionality. Crucially, group-mindedness is not the result of deliberation, but rather of a combination of naturally evolved sensitivities and culturally evolved practices of socialization. In Section 9.5, I will focus on these practices. I will characterize the socio-culturally scaffolded development of group-mindedness in ontogeny as a kind of "mindshaping" – a notion that is derived from the philosophical debate on social cognition. In Section 9.6, I will argue that intergenerational cycles of mindshaping form self-organizing processes that are an integral part of the explanation of how human beings manage to survive in large cultural groups. In Section 9.7, I will speculate about lessons for the philosophical debate on collective intentionality by suggesting a possible analogy for mindshaping group-mindedness in the domain of companies and cultural institutions.

9.2 Collective Intentionality and "Constitutive Top-Down Influence"

The term "collective intentionality" was introduced by Searle in 1990, but the concept it designates has a longer history (Searle, 1990). Durkheim's notion of "collective consciousness" and Collingwood's "joint enterprise", for example, already thematize phenomena that largely overlap with what Searle sought to make explicit (Collingwood, 1947; Durkheim, 1893/1984). Like most later theorists of collective intentionality, Searle starts from simple, small-scale examples. When you and I jointly intend to take a walk together, paint a house together, or go on vacation together, we do more than merely intend to walk, paint, or go on vacation at the same time and place; we intend to do these things *as* a team. We have, as some philosophers put it, "we-intentions".

There is quite general agreement over the fact that we-intentions are not aggregates of in "I-intentions" (intentions pertaining to actions of the intender only). But it is also generally held that intentions can only be had by individuals. This gives rise to the problem that drives the debate. Intentions require minds, and though some philosophers allow for group minds (Budelmann, 2017; Huebner, 2014; Theiner, 2013, 2014; Theiner et al., 2010), most philosophers think that group minds exist – at best – only metaphorically. Metaphorical entities cannot have real intentions, so we-intentions are had by individuals. But how can we conceive of that? How can *I* intend for *us* to do something? Answering this question contributes to explaining how there can be things like cultural institutions, according to Searle, because institutions hinge on collective intentions: treating slips of paper as money, for example, is collectively intending to treat these slips as having specific status functions, according to him.

The philosophical debate on collective intentionality has resulted in a variety of solutions to the problem. These range from thoroughly individualistic and non-normative to fully anti-individualist and normatively rich. An example of the first type of solution is Michael Bratman's view of collective intentionality based on common knowledge of each other's intentions and meshing sub-plans (Bratman, 2014). An example of the latter type is Margaret Gilbert's idea that collective intentionality is grounded in joint commitments resulting in an irreducible plural subject (Gilbert, 2014). This is not the place to discuss the range of options, though.

What must be discussed as background for the argument that follows is the fact that no philosophical theory of collective intentionality involves what I will call "constitutive top-down influence" of group-level processes on the decision, intentions, or actions of individual members of groups. The term "constitutive" is important here. For it is widely agreed that group-level processes can and do play a role in the actions and considerations of individuals that jointly make up groups to which intentions can be ascribed. Individual compliance with collective decision-making procedures, for example, is a case in point (List & Pettit, 2011). And so is commitment to a group and its goals (Gilbert, 2014). We can also think of peer pressure, or the fact that individuals adjust their we-intentions on the basis of perceived changes to a group's situation (think of football players of the same team silently changing their we-intentions to start defending due to developments in the game). What such cases have in common is that group-level processes influence the formation of we-intentions in individuals because they are used as inputs for the normal individual cognitive processes that underlie we-intention formation. We might call this "rational" or "cognitive" top-down influence. What I mean by "constitutive" top-down influence is something much more fundamental, namely influence of group-level events on the very *nature* of individual processes of we-intention formation. Group-level events, in this case, do not influence we-intention formation by co-determining *what* an individual decides or intends, but by co-determining *how* an individual decides and intends.

In her work on cultural evolution, Cecilia Heyes uses the metaphor of "grist" and "mills" (Heyes, 2018). "Grist" stands for the information that our cognitive systems – our minds – process. "Mills" stands for the cognitive (sub-)systems themselves. This metaphor helps to explain the difference between cognitive and constitutive top-down influence. Cognitive top-down influence is at play when group-level events affect the grist that is used to form we-intentions. Constitutive top-down influence is at play when group-level processes affect the mills behind the formation of collective intentions.

Cognitive top-down influence is very much present in the philosophical debate on collective intentionality. But constitutive top-down influence is completely absent from it. This is directly connected with a general weariness of metaphysically awkward notions such as "hive minds", or outdated ideas about historical teleology and *zeitgeist*. I completely share this weariness. But I want to argue, nevertheless, that constitutive top-down influence in collective intentionality is a real phenomenon. It is absent from the philosophical debate because the *formation* of individuals as more-or-less autonomous decision-makers and intenders is not part of it. Individuals figure simply as fully formed self-governing agents. I want to argue that by taking an evolutionary perspective on collective intentionality, the formation of individuals as members of groups that can be characterized by collective intentionality becomes part of the picture. And in this formative evolutionary process, there is ample room for constitutive top-down influence – without any metaphysical mystery.

9.3 The Shared Intentionality Hypothesis

If chimpanzees and bonobos are a clue to what our remote ancestors were like, we must conclude that in six or seven million years (since our last common ancestor), we have evolved from groups in which collaboration is rudimentary at best to ultra-social beings able to collaborate in very large groups with massively complex division of labor and cultural institutions. To explain this kind of evolutionary leap, a scalar notion such as collective intentionality – that can be applied to dyads as well as to large groups – is eminently suitable. Tomasello and colleagues capitalize on this scalar nature by distinguishing small-scale and large-scale collective intentionality as similar but different stages in the development toward increasingly more encompassing collaboration. They are not so much interested in *theories* of we-intentionality, but rather in the *phenomenon* itself, the different scales at which it is manifested, and the differences in psychological capacities required for various scales. They label small manifestations like dyadic collaboration "joint intentionality" and reserve the term "collective intentionality" for large-scale collaborations. Joint intentionality and collective intentionality are both manifestations of what they call "shared intentionality". The central evolutionary thesis – the "shared intentionality hypothesis" – contends that:

> the uniquely human adaptations for cooperation evolved in two key steps. The first step comprise adaptations enabling early human individuals to cooperate with one another dyadically in obligate collaborative foraging (with partner choice); these are the skills and motivations of joint intentionality. The second step comprised adaptations enabling modern human individuals to cooperate with one another in the larger collaborative enterprise known as culture; these are the skills and motivations of collective intentionality.
>
> (Tomasello, 2019b: 9)

What Tomasello and colleagues add to the philosophical debate is an emphasis on psychological requirements for the various scales of shared intentionality. This produces a more marked difference between small- and large-scale shared intentionality than the difference between scales of collective intentionality in the philosophical debate. The "skills and motivations" required for joint intentionality include capacities such as mind-reading, gaze tracking, and joint attention. The skills and motivations required for collective intentionality include capacities for cultural learning and referential communication.

Despite the differences in psychological capacities required for them, joint and collective intentionality are very similar phenomena. Both are characterized by what Tomasello calls "we > me [we-over-me] self-regulation". The "we" is the collaborative unit to which collaborators have committed themselves and which they must respect to retain their identity and credibility as collaborators (see Geurts' chapter in this volume for a philosophical analysis of such commitment). It is internalized by collaborators and drives their actions in ways that exceed simple strategic considerations. Of course, collaborators expect to get something out of the collaboration that individuals themselves would not be able to obtain, and of course, there is thus a self-directed motivation in joining a collaboration. But once entered into it, the "we" drives their actions in a *normative* way by giving rise to mutually felt obligations. These are not based on sympathy for collaboration partner(s), according to Tomasello, but rather on an understanding that (i) all participants in a collaboration are causally required for the desired outcome, (ii) partners can switch roles when needed, and (iii) the standards of performance for these roles are impersonal (Tomasello, 2019b: 6).

Because joint and collective intentionality both involve we > me self-regulation, Tomasello joins the philosophers in treating small-scale collaboration as a model for large-scale collaboration (with the explicit understanding that the psychological mechanisms involved in collective intentionality are different and much more complex than those involved in joint intentionality). After describing joint intentionality in simple dyadic collaborations between infants in terms of we > me self-regulation based on felt obligations, he describes large-scale collaboration as a scaled-up version of basically the same phenomenon: "In [a] cultural-institutional context . . . individuals continue to self-regulate in a we > me manner, but now the "we" is our culture" (Tomasello, 2019b: 7). This is because

> a cultural group . . . is nothing more or less than one big collaborative activity in which "we" as a people operate with a collective commitment to the group's surviving and thriving. Each individual has her role(s) to play in this collective commitment – both as a member of the group in general and, possibly, as a person playing some more specific division-of-labor role – and this generates, in a scaled-up manner, more universal normative expectations.
>
> (Tomasello, 2019b: 7)

One of the main theoretical perks of this modeling is that it allows for a theory of the origins of moral obligation in large groups by extrapolating from the kind of normativity that emerges in dyads. In dyads, this normativity can motivate protest in one of the collaborators when the other collaborator does not put sufficient effort into the joint project. Moral and legal regulations in societies are basically extrapolated versions of this: "The enforcement of social norms is . . . a kind of scaled-up, third-party version of the second-personal protest characteristic of joint intentional collaboration" (Tomasello, 2019b: 7). It would require a more elaborate exercise to derive political, economic, religious, and other institutions from joint intentionality in this way, but Tomasello can simply join the philosophers of collective intentionality (most notably Searle, 1990, 2010), who use that notion to explain institutional reality.

Joint and collective intentionality characterize stages in the evolution of humans as an increasingly more collaborative species. But because phylogeny is often reflected in ontogeny, these stages can also be studied in the development and socialization of human infants, where sensitivity for collective intentionality sets in early:

> [C]hildren from about three years of age (but, needless to say, not apes) socially self-regulate on the basis of cultural structures – such as, prototypically, conventional and moral norms – that are based in cognitive

> processes of collective intentionality, what we may call normative self-governance. This process operates in much the same way as social self-regulation in a dyadic context, but the social other in this case is more generalized and authoritative.
>
> (Tomasello, 2019a: 38)

Although collective intentionality "operates in much the same way" as joint intentionality, it does not automatically develop from dyadic collaborations. The internalization of "cultural structures" is a separate step, to which I will now turn.

9.4 Group-Mindedness

Tomasello describes we > me self-regulation in *collective* intentionality in different ways. Sometimes he describes it as if it were the result of rational internal deliberation. At other times, he describes it more as a non-reflective, automated attitude acquired through upbringing. Let me begin with the first characterization. I will argue that this should be interpreted metaphorically and that the second characterization is the more realistic depiction of how we become proficient collaborators in complex cultural groups.

Here is an example of the rational deliberation characterization of we > me self-regulation in collective intentionality:

> Because we all value the group's smooth functioning, we all must do things in the ways that we all expect us to do them. In addition, to be a good group member, we must also make sure that others follow these norms as well (especially by normatively protesting violations).
>
> (Tomasello, 2019b: 7)

It is not at all clear whether Tomasello intends the "because" at the start of this quote as a psychological "because" – a conscious reason for us to keep to the norms of our society – but it is at least possible to read the quote as such. I believe we should not take this reading literally. For it is quite implausible that when adhering to the rules and norms of our society, most of us are consciously aware of such reasoning.

Most human societies are incredibly complex networks of roles – determined by economic division of labor and cultural institutions – that interlock and together secure the survival and thriving of the group. The effect of an individual's actions on the overall group functioning is impossible to fathom and, in isolation, very probably negligible in most instances. It would be a clear overintellectualization to think that each time we keep to the norms and regulations of our society, we do so because we are motivated by an awareness of making a contribution to the "smooth functioning" of society. It sounds more reasonable to say that, in general, we decided at some point that keeping to the norms and rules of society is a good thing, after which we (*ceteris paribus*) stick to the corresponding courses of action. But even this is probably an overintellectualization. Was there ever such a "decision"? Or is that a rational reconstruction of the processes of socialization through which we have been initiated in these norms and rules? If we look at the phenomena of conformity bias,[2] over-imitation, natural pedagogy, and the eagerness with which infants absorb and apply social rules (Csibra & Gergely, 2006; Horner & Whiten, 2005; Nielsen & Tomaselli, 2010; Rakoczy & Schmidt, 2013; Rakoczy et al., 2008; Schmidt et al., 2012; Stengelin et al., 2019), it seems overwhelmingly likely that we never really decided to adopt the norms and rules of our society, but that we adopted these gradually by resonating with our social environment.

But this rejection of the deliberative reading of we > me self-regulation may be too quick, for Tomasello also stresses that this regulation is driven, at least in part, by anticipating the normative judgments of others (e.g., Tomasello, 2019a: 37). "I keep to the rules we share because I do not wish others to think poorly of me" – this may seem like a form of deliberation. I think there are problems with such a deliberative reconstruction, however.

We should keep in mind that incurring negative normative judgments is experienced as a kind of sanction only because we have a wish to avoid such judgments, and this wish is itself *not* the product of deliberation. We have a natural tendency to mind the judgments of others, which is completely unreflective. The evolution of this tendency may well have a strategic explanation (minding the judgments of others will likely make others act nicer toward me), but that does not mean that the tendency itself is sustained by strategic reasoning. Hence, even when we describe we > me self-regulation as a form of deliberation aimed at avoiding negative judgments of others, there is still a crucial non-reflective component in it.

Something similar applies to the other side of the equation, the tendency of us to judge and, if necessary, sanction trespassers. Humans distinguish themselves from non-human animals by engaging in what is known as "third-party punishment" – punishing trespassers on behalf of the group and not as a form of personal retaliation. Third-party punishment is an important ingredient of Tomasello-style collective intentionality. But it cannot be the result of rational deliberation, for punishing is costly and in third-party punishment there are no direct returns to be expected by the punisher. Why do we do it nevertheless? Why do we bother to approve or disapprove of someone else's behavior when this was not directed at us (that we *do* bother is quite evident from our thirst for scandals, our unstoppable tendency to gossip, and frequent public calls for "justice" in case of serious crimes). The very possibility of anticipating social sanctions hinges on an evolved non-strategic tendency, then, similar to the tendency to overimitate, adopt rules, be instructed by our caregiver's gaze, and so on (see, e.g., Henrich & Muthukrishna, 2021).

All in all, we > me self-regulation in collective intentionality must be considered as being the product of pro-social, conformist, and conventional tendencies in us that have good evolutionary explanations.[3] We *self*-regulate because *we* adjust our own behavior in accordance with the rules and norms of our environment. But the motivation for doing so is not our individual rational or strategic judgment. Rather, it is a set of mental and behavioral dispositions that is ingrained in us irrespective of rational considerations. This unreflective set of dispositions is what Tomasello calls "group-mindedness". His use of this term supports the second reading of we > me self-regulation as a non-reflective automated process. Group-mindedness is described as a psychological phenomenon that is the result of a combination of innate tendencies and cultural upbringing. It gradually matures during socialization. It can be *reconstructed* as an internal deliberation aimed at the smooth functioning of the group. Such a reconstruction can make us understand why this attitude is selected. But it does not describe a psychological reality (Slors, 2025).

The psychological reality behind group-mindedness is described by Tomasello as a process of "maturation of capacities for collective intentionality" (Tomasello, 2019a: 273). This is tantamount to the internalization of a group's norms and rules as a kind of cultural "über-ich" (Slors, 2024). Group-mindedness starts to operate around the age of 3:

> When three-year-olds start enforcing social norms, including on third parties from a noninvolved position, this can have an element of social learning and conformity . . .: they are doing what adults do when they enforce norms. But, unlike two-year-olds, three-year-olds do not need an adult model: they infer a generic norm from a single observation of an intentional action; they enforce that norm with no input from adults; and they use normative language in their enforcement. These facts suggest that three-year-olds are indeed coming to understand the group-minded nature of social norms and how they work. The fact that children enforce conventional norms more strongly on in-group than out-group members also suggests a group-minded understanding of norms.
>
> (Tomasello, 2019a: 271–272)

Even though Tomasello presents group-mindedness as an innate tendency, it is important for what follows to emphasize two provisos. First, as the phenomenon of feral children shows, without the presence of a socio-cultural environment in which norms and rules function and are enforced, group-mindedness does *not* emerge (Chambers, 2002; McNeil et al., 1984). The maturation Tomasello speaks of is thus not something akin to the ripening of fruit; it is something that happens only in a socio-cultural context through exchanges, interactions, and socialization. Secondly, the norms and rules that infants come to employ – apart from self-invented rules of games – are highly culture specific.

9.5 Mindshaping

The development of group-mindedness as an unreflective set of mental and behavioral dispositions can be described as a form of "mindshaping" – the cultural acquisition of specific cognitive mills in Heyes' metaphor (see Section 9.1). I borrow the term "mindshaping" from the debate on folk-psychology (a.k.a. theory of mind or mindreading), so I must start this section with a brief digression into an adjacent field of inquiry.

Folk-psychology is the practice of explaining, understanding, and predicting the behavior of others by imputing mental states such as beliefs and desires. The term "mindshaping", which was coined as early as 2001 (Mameli, 2001), was influentially put on the philosophical agenda by Tad Zawidzki (Zawidzki, 2008, 2013). Zawidzki argues, briefly put, that to explain behavior of others in terms of folk-psychology is cognitively intractable when all logically possible options for such an explanation are on the table. A single action can logically be interpreted as being motivated by many different sets of mental states, and figuring out which of these is the correct one requires taking into account all kinds of background assumptions and contextual data. Logically speaking, folk-psychology is thus tremendously unwieldy. In practice, however, we effortlessly apply folk-psychological explanations to the actions of others. How do we do that? Zawidzki argues that in the course of our upbringing, we learn to behave according to specific culturally sanctioned subset of folk-psychological options. We learn to regiment our behavior to make it "readable" by others. This learning does not itself depend on a full-blown ability to mindread, but rather on a more rudimentary form of social cognition.

In a series of articles, Victoria McGeer has enriched this idea of mindshaping by describing folk-psychology as a normative practice and by embedding the process of mindshaping in the wider literature on the enculturation of minds (McGeer, 2007, 2015, 2021). This version of the idea of mindshaping extends beyond folk-psychology to other kinds of normative practices (Slors & Kiverstein, 2025). McGeer takes her lead from Clifford Geertz's famous quote:

> Man's nervous system does not merely enable him to acquire culture; it positively demands that he do so if he is going to function at all. Rather than culture acting only to supplement, develop and extend organically

> based capacities logically and genetically prior to it, it would seem to be ingredient to those capacities themselves.
>
> (Geertz, 1973: 68)

It becomes more and more clear that in the first two decades of our lives, the innate cognitive abilities of our brains are being augmented and transformed through exposure to and internalization of cultural practices (Wexler, 2006). Neural adaptations such as the "visual word form area" that develops in response to reading and writing practices (Dehaene et al., 2015) are well known, but cultural anthropological and developmental psychological evidence suggests that similar adaptations occur in connection with very many other cultural practices as well (Henrich, 2016, 2020; Heyes, 2018).

McGeer's claim is that the use of folk-psychology is such a practice. Internalizing the rules and regularities of this practice means shaping our cognitive apparatus to yield a near[4] automatized practical know-how without which it is impossible to navigate the social worlds of the societies we live in. This is largely in line with Heyes' views on the development of a "cognitive gadget" devoted to mindreading.[5] What is important is that the mindshaping approach does not stop with folk-psychology. McGeer's conception of folk-psychology is already much broader than Zawidzki's. Mindshaping, on her account, also involves the acquisition of proficiency in all kinds of cultural conventions and customs. What connects these practices is that they allow us to coordinate our behavior in order to collaborate efficiently.

The rules and norms of the society one grows up in are, like folk-psychology, used to coordinate collaborative behavior. By internalizing and (near-)automatizing these, individuals become reliable and predictable collaborators. Coming to understand how to behave like a group-minded human being in Tomasello's sense is coming to internalize norms and rules as near-automatic default behavior. This is likely to involve augmenting and modifying innate neural circuitry in more or less the same way that learning to read leads to the formation of a visual word form area. The scaffolded development of group-mindedness, in other words, can be regarded as a form of mindshaping too.

There need not be a contradiction between Tomasello's insistence on the innateness of group-mindedness and the idea that this must be developed through mindshaping. For it may well be that our susceptibility to this type of mindshaping is innate. In fact, that is exactly how Tomasello interprets experiments on overimitation and rule-adoption by young infants.

9.6 Self-Organizing Cycles of Mindshaping

The idea that group-mindedness is the result of mindshaping is a necessary but not a clearly sufficient step in arguing for the idea that collective intentionality, Tomasello-style, involves constitutive top-down influence. Mindshaping, if anything, is co-constituting the minds and hence the intentions and actions of group members. But it is not yet clear why this constituting must be thought of as a top-down process in which group-level events determine how individuals function. For it may be thought that *individual* caregivers may intend infants to become group-minded. And as the quote at the end of Section 9.3 shows, mindshaping group-mindedness may well consist largely of *individual* infants imitating others on their own initiative. Where is the top-down influence?

I argue that this influence becomes visible only when we zoom out and consider mindshaping group-mindedness as a process that encompasses many generations. Once we are mindshaped into group-minded beings, this means not only that we have internalized the norms and rules of our community, but also that we tend to enforce these in others, even in instances where we have no skin in the game. This is not only a tendency we see in young infants. We retain this tendency throughout our lives, and we put it to use when raising our own offspring. Once we are mindshaped as group-minded beings, we have become mindshapers ourselves. This turns mindshaping into a trans-generational cyclical process with its own supra-personal, intergenerational dynamics – that is, a group-level process.

No one is in charge or in control of this process, because we only become agents who are capable of being in control or in charge after we are socialized – but then mindshaping has already exerted its influence. Our susceptibility to being shaped into group-minded beings is not the result of anyone's decision or intention, but of blind evolutionary processes that have selected groups with good collaborators. That is the importance of the realistic reading of Tomasello's notion of collective intentionality and we > me self-regulation defended in Section 9.3. We can rationally reconstruct our adherence to the norms and rules of our culture as motivated by a desire for the smooth functioning of the group. And we can make our practices of mindshaping group-mindedness intelligible by depicting them as being motivated by a (collective) intention to properly raise our offspring. But only by bracketing the fact that we – the mindshapers – are ourselves the products of mindshaping and that this desire and these intentions are themselves the products of blind evolutionary processes. But while we can bracket this fact for the sake of explaining a given practice, we cannot ignore it. The cycle of generations that mindshapes the next generation into mindshapers that promote group-mindedness, etc., is a self-organizing process that is driven by natural selection and not by anyone's intention.

That, I believe, is the anti-individualistic consequence of Tomasello's adaptation of the notion of collective intentionality. In this picture, there is a clear constitutive top-down influence of group-level processes on the individual contributions to collective intentionality. But this does not mean that Tomasello or other thinkers from the domain of cultural evolution buy into a mysterious hive mind or any other metaphysically objectionable notion. There is nothing mysterious about the natural selection of practices that turn out to have beneficial effects on the survival and thriving of groups.

9.7 From Cultural Evolution to Social Ontology

There are several ways in which Tomasello's use of the notion of "collective intentionality" differs from the use of this term in philosophy. Tomasello does not attempt to solve the puzzle of we-intentions and focuses on the psychological requirements of joint and collective intentionality instead. These differences, however, do not stand in the way of the possibility that we can apply insights from the use of the notion of collective intentionality in the context of cultural evolution in the domain of the philosophy of social ontology. In this last section, I will concentrate on the possibility of applying the non-mysterious kind of top-down influence that is implicit in Tomasello's view outlined in the previous sections: can a similar kind of non-mysterious top-down influence exist in, say, institutions or corporations (see the chapter by Stoelhorst in this volume)? If so, we must find a parallel to the practices of mindshaping group-mindedness. We must also ground these possible parallels in our evolved psychology. Let me say something about both requirements, starting with the first.

Sociologists distinguish between primary and secondary socialization. Primary socialization is the acquisition of basic norms, values, and modes of behavior in the context of child-rearing by family and close relatives. Secondary socialization refers to the further initiation of infants and adults in the ways of a culture by people outside the family circle. It includes, but is not limited to, formal education and specialization in a certain profession (Zerubavel, 1997). The process of mindshaping group-mindedness largely overlaps with primary socialization. Becoming initiated in the ways and organization of, say, a company is typically a form of secondary socialization in which one comes to internalize this particular company's rules, norms, and practices. The question is to what extent secondary socialization is sufficiently like primary socialization to think of it as a mechanism for top-down influence of group-level processes (the company as a whole) on individual contributions to the collective intentionality involved in the company's "actions" and "intentions".

The first thing to note here is that there are crucial differences between primary and secondary socialization that directly affect the parallel. Mindshaping group-mindedness pertains to infants who unreflectively follow, imitate, listen to, and are corrected by adult models. Familiarizing a new recruit with the rules, norms, and customs of a given company, by contrast, may seem to involve an autonomous, reflective, rational adult being instructed by one or more other autonomous, reflective, rational adults. We tend to think that the new recruit adopts the rules and norms of the company through reflective understanding, not through (over)imitation. If this picture is correct, the recruit's initiation is not like mindshaping group-mindedness. But a depiction of the recruit's initiation as a fully reflectively controlled process is very likely an overintellectualization. The rules and practices of a company are just as much a normative practice as folk psychology or the rules and norms of a society. And initiation into such a practice is very likely to involve imitation and implicit or explicit correction by others. Experts teach the company's normative practice to novices, but certainly not only through explicit instruction. It is highly likely that by the time a novice transitions into a seasoned company employee, she will have internalized and automatized many of the company's practices through a form of natural pedagogy.[6]

The main question to be answered, however, is whether the perpetuation of the company's rules and norms can be considered a self-organizing group-level phenomenon, or whether it can be reduced to an aggregate of individuals' intentions (including we-intentions). To answer this question, we must be able to assess the extent to which unreflective automatization of rules and norms is an essential ingredient of the continuation of the company. Is it realistic to think that, in principle, a company could survive and thrive purely on the basis of the conscious intentions of its employees? Opinions might differ on this. But at least two considerations should incline us toward a negative answer. The first of these is that there is by now massive evidence for the idea that humans conduct a great deal of their behavior in an automated, unreflective way. Reference to the movement in social psychology under the heading of "the new unconscious" (Hassin et al., 2005) and the widespread application of dual processing accounts of human cognition (Evans, 2008) should suffice to make this point.

The second consideration brings us to the second requirement for the idea that there is a parallel between primary and secondary socialization that warrants the possibility of top-down influence in processes of collective intentionality as studied in the debate on social ontology. To what extent does our evolved psychology play a role in the collective intentionality that underlies the functioning of companies and institutions? Perhaps Tomasello is right in claiming that we have an innate tendency toward being mindshaped into group-mindedness. But does this imply we have an innate tendency toward, say, "company-mindedness"? Well, perhaps partly. Of course, the decision to join a company is usually not instinct-driven or out of our hands (in the way that being born into a community is out of our hands). But once we are part of a company, or a member of an institution, part of a circle of friends, etc., it is not at all strange to assume that our group-mindedness, which is the result of our evolved psychology in combination with primary socialization, comes into play. We are evolved to be "socializable" into good and reliable collaborators. There is no reason whatsoever to assume this psychology will cease to exert its influence on our behavior in the setting of, for example, a company.

So, all in all, it is likely that at least a weak parallel to the kind of top-down influence that I have argued is implicit in Tomasello's adaptation of the notion of collective intentionality can also be found in the context of companies, institutions, and other collectives that are studied by social ontologists. To what degree such top-down influence plays a role is hard to say and is very probably different in each situation. But the very possibility that it is there – as a subject for future research – is something that the debate on social ontology can learn from the debate on cultural evolution – by treating humans not simply as autonomous, rational decision-makers, but as decision-makers who are endowed with motivations and tendencies that are the product of cultural evolution and who are shaped by processes of primary and secondary socialization.

9.8 Summary and Conclusion

Constitutive top-down influence of group-level processes on the actions and intentions of individuals involved in group-level processes that can be described in terms of collective intentionality is completely absent from the philosophical debate on this topic. In the philosophical debate, individuals are treated as fully formed, mature decision-makers who can take group-level processes into account in their decisions and intentions, but only by using them as inputs for the normal cognitive processes that underlie such mental states. Constitutive top-down influence is associated with metaphysically awkward theories.

The adaptation of the notion of collective intentionality in the debate on cultural evolution, in particular by Michael Tomasello and colleagues, paints a different picture. The acquisition of our capacities for collective intentionality, as a crucial stage in our evolutionary phylogeny, involves a kind of group-mindedness that cannot be accounted for in terms of individual rational considerations. Instead, it must be accounted for in terms of group-level cycles of socialization that can be regarded as a kind of mindshaping that builds on evolved tendencies for overimitation, norm-sensitivity, and conformity bias. In the evolutionary picture of collective intentionality, the *formation* of individuals as full-blown decision-makers and intentional agents is part of the picture too. And this formation involves the constitutive influence of group-level processes – in the form of trans-generational cycles of mindshaping that build on an evolved susceptibility for this – on the intentions and decisions of individuals that, in aggregated form, can be described as displaying collective intentionality. But this constitutive influence is not of a metaphysically objectionable kind.

Thus, by adapting the notion of collective intentionality, the debate on cultural evolution adds a dimension of influence that is so far absent from the philosophical debate. The philosophical debate focuses not so much on human cultural groups, but rather on entities such as companies, organizations, and institutions. In the last section of this chapter, I have argued that the evolutionary picture might be put to use in the philosophical debate as well. Our evolved susceptibility to mindshaping group-mindedness is likely to facilitate the initiation of individuals into the rules and norms of organizations, companies, and institutions. Thus, we might view such initiation as stages in trans-generational cycles of initiation that sustain these companies, organizations, and institutions. These group-level cycles exert constitutive top-down influence on individuals who jointly enact the processes of companies, etc., that can be described in terms of collective intentionality.[7]

Notes

1 I am using this term in a very broad sense here to refer to debates on the origins of human culture and the way in which culture as an adaptation has shaped our minds. These debates include the more focused discussion on selectionist and "populationist" (Heyes, 2018) or "kinetic" (Lewens, 2015) approaches to the evolution of practices, ideas, and (according to Heyes) cognitive abilities within cultures. But they also include the study of the evolutionary development of human collaboration in cultural groups as such.

2 I am not concerned here with high-profile and controversial (Padalia, 2014) experiments purportedly showing our inclination toward extreme or bizarre obedience (e.g., Asch, 1956; Milgram, 1963; Zimbardo, 1973). The kind of conformity I am referring to is the everyday phenomenon that people tend to adopt the customs, rules, conventions, beliefs, and norms of the social environment they grow up.

3 These explanations differ somewhat between authors, but gene–culture co-evolution (Henrich, 2016, 2020; Henrich & Boyd, 1998; Muthukrishna et al., 2016; Richerson & Boyd, 2005; Richerson et al., 2010) and multi-level selection (Smaldino, 2014; Sober & Wilson, 1998; Wilson, 2003) are likely components.

4 In line with, for example, Christensen et al. (2016), McGeer argues that the justified backlash against an overly intellectualized picture of our dealings with the social world in terms of fully automatic and mindless responses (Dreyfus, 1997) is not in line with the way in which these responses are usually *intelligent,* which requires what she calls "mindful embodied expertise".

5 This view of our mindreading abilities as culturally acquired seems at odds with the way Tomasello thinks about it as a more or less naturally given precondition for joint intentionality (Heyes & Moore, 2023). Like Zawidzki, however, McGeer does acknowledge that the acquisition of folk-psychological proficiency requires infants to understand the instructions and intentions of their caregivers. This more rudimentary form of social cognition may also suffice for joint intentionality. At any rate, mindreading is not the topic of concern here – the main issue is that the scaffolded development of group-mindedness is a form of mindshaping.

6 Economists Akerlof and Kranton describe such processes in terms of a change in an individual's preferences, that is, a change in a person's identity (Akerlof & Kranton, 2005). The terminology they use is different from the terminology used in this chapter, but their descriptions and examples perfectly match the processes of initiation I have in mind. Crucially, these involve all kinds of non-reflective processes that make use of our evolved pro-social nature.

7 Thanks to the editors of this volume, Bart Geurts, and especially to Jan-Willem Stoelhorst, for very helpful comments and suggestions for this chapter.

References

Akerlof, G.A. & Kranton, R.A. (2005): Identity in the economics of organizations, *Journal of Economis Perspectives* 19: 9–32.

Angelino, L. (2022): (Re)conceptualizing the genesis of a 'we is greater than me' psychological orientation: Sartre meets Tomasello, *Journal of Social Ontology* 8: 68–93.

Angus, S.D. & Newton, J. (2015): Emergence of shared intentionality is coupled to the advance of cumulative culture, *PLoS Computational Biology* 11: e1004587.

Asch, S.E. (1956): Studies of independence and conformity: I. A minority of one against a unanimous majority, *Psychological Monographs: General and Applied* 70: 1–70.

Boehm, C. (2018): Collective intentionality: A basic and early component of moral evolution, *Philosophical Psychology* 31: 680–702.

Bratman, M.E. (2014): *Shared Agency: A Planning Theory of Acting Together*, New York: Oxford University Press.

Budelmann, F. (2017): Group minds in classical Athens? Chorus and dēmos as case studies of collective cognition, in: Anderson, M. & Cairns, D. (Eds): *Distributed Cognition in Classical Antiquity*, Edinburgh: Edinburgh University Press, pp. 194–212.

Chambers, T.L. (2002): Feral children, *Journal of the Royal Society of Medicine* 95: 429–430.

Christensen, W., Sutton, J. & McIlwain, D.J.F. (2016): Cognition in skilled action: Meshed control and the varieties of skill experience, *Mind & Language* 31: 37–66.

Collingwood, R.G. (1947): *The New Leviathan: Or: Man, Society, Civilization and Barbarism*, Oxford: Clarendon Press.

Csibra, G. & Gergely, G. (2006): Social learning and social cognition: The case for pedagogy, in: Munakata, Y. & Johnson, M.J. (Eds): *Processes of Brain and Cognitive Development: Attention and Performance XXI*, Oxford: Oxford University Press, pp. 240–274.

Dehaene, S., Cohen, L., Morais, J. & Kolinsky, R. (2015): Illiterate to literate: Behavioural and cerebral changes induced by reading acquisition, *Nature Reviews Neuroscience* 16: 234–244.

Dreyfus, H.L. (1997): Intuitive, deliberative, and calculative models of expert performance, in: Zsambok, C.E. (Ed): *Naturalistic Decision Making*, Mahwah, NJ: Laurence Erlbaum, pp. 17–28.

Durkheim, E. (1893/1984): *The Division of Labour in Society* (W.D. Halls, Transl), London: Macmillan.

Evans, J.S. (2008): Dual-processing accounts of reasoning, judgment, and social cognition, *Annual Review of Psychology* 59: 255–278.

Geertz, C. (1973): *The Interpretation of Cultures*, New York: Basic Books.

Gilbert, M. (2014): *Joint Commitment: How We Make the Social World*, New York: Oxford University Press.
Hassin, R.R., Uleman, J.S. & Bargh, J.A. (2005): *The New Unconscious*, New York: Oxford University Press.
Henrich, J. (2016): *The Secret of Our Success: How Culture Is Driving Human Evolution, Domesticating Our Species, and Making Us Smarter*, Princeton: Princeton University Press.
Henrich, J. (2020): *The Weirdest People in the World: How the West Became Psychologically Peculiar and Particularly Prosperous*, New York: Farrar, Straus and Giroux.
Henrich, J. & Boyd, R. (1998): The evolution of conformist transmission and the emergence of between-group differences, *Evolution and Human Behavior* 19: 215–241.
Henrich, J. & Muthukrishna, M. (2021): The origins and psychology of human cooperation, *Annual Review of Psychology* 72: 207–240.
Herrmann, E., Call, J., Hernàndez-Lloreda, M.V., Hare, B. & Tomasello, M. (2007): Humans have evolved specialized skills of social cognition: The cultural intelligence hypothesis, *Science* 317: 1360–1366.
Heyes, C. (2018): *Cognitive Gadgets: The Cultural Evolution of Thinking*, Cambridge, MA: Harvard University Press.
Heyes, C. & Moore, R. (2023): Henrich, Heyes, and Tomasello on the cognitive foundations of cultural evolution, in: Tehrani, J.T., Kendal, J. & Kendal, R.L. (Eds): *The Oxford Handbook of Cultural Evolution*, New York: Oxford University Press, pp. 230–240.
Horner, V. & Whiten, A. (2005): Causal knowledge and imitation/emulation switching in chimpanzees (*Pan troglodytes*) and Children (*Homo sapiens*), *Animal Cognition* 8: 164–181.
Huebner, B. (2014): *Macrocognition: A Theory of Collective Minds and Distributed Intentionality*, New York: Oxford University Press.
Lewens, T. (2015): *Cultural Evolution: Conceptual Challenges*, Oxford: Oxford University Press.
List, C. & Pettit, P. (2011): *Group Agency: The Possibility, Design, and Status of Corporate Agents*, Oxford: Oxford University Press.
Mameli, M. (2001): Mindreading, mindshaping, and evolution, *Biology and Philosophy* 16: 595–626.
McGeer, V. (2007): The regulative dimension of folk-psychology, in: Hutto, D.D. & Ratcliffe, M. (Eds): *Folk-Psychology Re-Assessed*, Dordrecht: Springer, pp. 137–156.
McGeer, V. (2015): Mind-making practices: The social infrastructure of self-knowing agency and responsibility, *Philosophical Explorations* 18: 259–281.
McGeer, V. (2021): Enculturating folk psychologists, *Synthese* 199: 1039–1063.
McNeil, M.C., Polloway, E.A. & Smith, D.J. (1984): Feral and isolated children: Historical review and analysis, *Education and Training of the Mentally Retarded* 19: 70–79.
Milgram, S. (1963): Behavioral study of obedience, *The Journal of Abnormal and Social Psychology* 67: 371–378.
Muthukrishna, M., Morgan, T.J.H. & Henrich, J. (2016): The when and who of social learning and conformist transmission, *Evolution and Human Behavior* 37: 10–20.
Nielsen, M. & Tomaselli, K. (2010): Overimitation in Kalahari bushman children and the origins of human cultural cognition, *Psychological Science* 21: 729–736.

Padalia, D. (2014): Conformity bias: A fact or an experimental artefact?, *Philosophical Studies* 59: 223–230.

Pettit, P. (2020): A hard choice for Tomasello, *Behavioral and Brain Sciences* 43: e81.

Rakoczy, H. (2009): Collective intentionality and the roots of human societal life, in: Röska-Hardy, L.S. & Neuman-Held, E.M. (Eds): *Learning from Animals? Examining the Nature of Human Uniqueness*, London: Psychology Press, pp. 105–121.

Rakoczy, H. & Schmidt, M.F.H. (2013): The early ontogeny of social norms, *Child Development Perspectives* 7: 17–21.

Rakoczy, H., Warneken, F. & Tomasello, M. (2008): The sources of normativity: Young children's awareness of the normative structure of games, *Developmental Psychology* 44: 875–881.

Richerson, P.J. & Boyd, R. (2005): *Not by Genes Alone: How Culture Transformed Human Evolution*, Chicago: University of Chicago Press.

Richerson, P.J., Boyd, R. & Henrich, J. (2010): Gene-culture coevolution in the age of genomics, *Proceedings of the National Academy of Sciences* 107: 8985–8992.

Schmidt, M.F.H., Rakoczy, H. & Tomasello, M. (2012): Young children enforce social norms, *Current Directions in Psychological Science* 21: 232–236.

Searle, J. (1990): Collective intentions and actions, in: Cohen, P., Morgan, J. & Pollack, M.E. (Eds): *Intentions in Communication*, Cambridge, MA: MIT Press, pp. 401–416.

Searle, J. (2010): *Making the Social World: The Structure of Human Civilization*, Oxford: Oxford University Press.

Slors, M. (2024): Explaining the cultural evolution of large-scale collaboration: Conventionality as an alternative for collective intentionality, *Review of Philosophy and Psychology* 15: 933–953.

Slors, M. (2025): Cultural groups as survival vehicles and the concept of symbiotic cognition, *Cogent Arts and Humanities* 12: 2513139.

Slors, M. & Kiverstein, J. (2025): Mindshaping, folk psychology, and cultural institutions, in: Zawizki, T. (Ed): *The Routledge Handbook of Mindshaping*, London: Routledge, pp. 83–96.

Smaldino, P.E. (2014): The cultural evolution of emergent group-level traits, *Behavioral and Brain Sciences* 37: 243–254.

Sober, E. & Wilson, D.S. (1998): *Unto Others: The Evolution and Psychology of Unselfish Behavior*, Cambridge, MA: Harvard University Press.

Stengelin, R., Hepah, R. & Haun, D.B.M. (2019): Being observed increases overimitation in three diverse cultures, *Developmental Psychology* 55: 2630–2636.

Theiner, G. (2013): Transactive memory systems: A mechanistic analysis of emergent group memory, *Review of Philosophy and Psychology* 4: 65–89.

Theiner, G. (2014): A beginner's guide to group minds, in: Sprevak, M. & Kallestrup, J. (Eds): *New Waves in Philosophy of Mind*, London: Palgrave Macmillan, pp. 301–322.

Theiner, G., Allen, C. & Goldstone, R.L. (2010): Recognizing group cognition, *Cognitive Systems Research* 11: 378–395.

Tomasello, M. (2014): *A Natural History of Human Thinking*, Cambridge, MA: Harvard University Press.

Tomasello, M. (2016): Precís of *a natural history of human thinking*, *Journal of Social Ontology* 2: 59–64.

Tomasello, M. (2019a): *Becoming Human: A Theory of Ontogeny*, Cambridge, MA: Harvard University Press.

Tomasello, M. (2019b): The moral psychology of obligation, *Behavioral and Brain Sciences* 43: e56.

Wexler, B. (2006): *Brain and Culture: Neurobiology, Ideology, and Social Change*, Cambridge, MA: MIT Press.

Wilson, D.S. (2003): *Darwin's Cathedral: Evolution, Religion, and the Nature of Society*, Chicago: University of Chicago Press.

Zawidzki, T.W. (2008): The function of folk psychology: Mind reading or mind shaping?, *Philosophical Explorations* 11: 193–210.

Zawidzki, T.W. (2013): *Mindshaping: A New Framework for Understanding Human Social Cognition*, Cambridge, MA: MIT Press.

Zerubavel, E. (1997): *Social Mindscapes*, Cambridge, MA: Harvard University Press.

Zimbardo, P.G. (1973): On the ethics of intervention in human psychological research: With special reference to the Stanford prison experiment, *Cognition* 2: 243–256.

10 Evolution and Social Ontology

The Modern Firm as a Major Transition in Cultural Evolution

J.W. Stoelhorst

10.1 Introduction

My main objective in this chapter is to answer the question "what is a firm?" This means that I will be answering an ontological question, in the sense that I will derive what I believe to be the essential features of modern firms. However, my approach to deriving these features is not primarily to engage in a philosophical exercise. It is rather to explain the modern firm, and the public corporation in particular, in terms of what an evolutionary perspective on human behavior and organization suggests are general insights into the nature of human social organization. In other words, my approach is to put science before philosophy: I build on and extend the evolutionary explanation of human behavior and organization to better understand the nature of a specific institution, the modern firm.[1] At the same time, in doing so, I also hope to contribute to the philosophical project of articulating a better understanding of social ontology more generally (cf. Hindriks & Guala, 2015; Lawson, 2012, 2015; Searle, 2005)

My answer to the question "what is a firm?" will be based on an extension of my previous work on how evolutionary theory can inform our understanding of human behavior and organization (in particular, Stoelhorst, 2005, 2008; Stoelhorst & Richerson, 2013). This work is informed by evolutionary epistemology (Campbell, 1960; Plotkin, 1994), gene–culture co-evolution theory (Henrich, 2004; Richerson & Boyd, 2005), the theory of multi-level selection (Sober & Wilson, 1998; Wilson et al., 2008), and the theory of major transitions in evolution (Maynard Smith & Szathmáry, 1995). Below, I will briefly summarize how a combination of these evolutionary theories offers a coherent framework for studying human social organization. I will then use this framework, and the concept of major transitions in evolution in particular, to analyze the nature of the modern firm. I will interpret the emergence of the corporate form as a major transition in cultural evolution and will discuss how an interpretation of the corporate form in these terms sheds light on the ontology of modern firms.

DOI: 10.4324/9781003643845-13

In addition to shedding light on the nature of the modern firm as one of the central institutions of our times, I hope to make two more general contributions. First, I show how the theory of major transitions in evolution, which was developed to account for the major transitions in biological evolution, can be generalized to cultural evolution. In doing so, I explicitly go beyond the traditional individual-level perspective of much of the cultural evolution literature (cf. Valković, 2024). Specifically, I will argue for a recognition of the corporation as a unit of selection with its own emergent properties that are irreducible to the properties of its individual members. Second, I show how understanding a particular modern institution (the corporation) as the product of a gene–culture co-evolutionary process sheds light on social ontology more generally. In particular, I will derive an understanding of human organization in terms of "roles" and "norms" as the linking pin between the individual and collective level of analysis, and an understanding of cultural evolutionary dynamics in terms of "coalitional competition". Recognizing the role of coalitional competition helps counter the criticism that theories of cultural evolution tend to sidestep questions of power (Fracchia & Lewontin, 1999).[2]

10.2 The Contested Ontology of the Corporation

Why focus on the firm as an object of inquiry in a paper on social ontology? One reason is that, supported by the cultural innovation of legal personhood, firms, and large public corporations in particular, are arguably the most influential institutions of our time. They have been and are, on the one hand, a fantastically successful cultural innovation that has allowed the human species to generate enormous economic wealth. On the other hand, they are also directly implicated in most of the so-called "grand challenges" that humanity faces today, including inequality, climate change, and the destruction of biodiversity (Davis, 2021). Moreover, in light of these challenges, there is an increasingly widespread sense that we are losing control of how firms are affecting us (e.g., Runciman, 2023). In other words, there are obvious practical reasons for wanting to better understand the nature of the modern firm.

Another reason to focus on the firm is that its existence raises interesting ontological questions. How are we to understand the nature of the firm in general and the corporation in particular? Is the firm merely a group of people, or is it an agent in its own right? Firms that are incorporated, that is, firms that adopt the legal form of a corporation, are recognized by law as separate legal persons. However, even among lawyers, there is no consensus about what this means for how we should interpret the nature of incorporated firms. There have been and continue to be at least three competing views of the corporation: the corporation as a concession, the corporation as a mere aggregate of individuals, and the corporation as a real entity (see, e.g., Winkler, 2018). Of these three views, the modern interpretation of the aggregate view is that the firm is a "legal fiction" and should be understood as a "nexus of contracts" among individuals (Jensen & Meckling, 1976). On this view, the firm is merely a group of people. In contrast, the real entity view emphasizes that incorporation leads to an entity with characteristics that are independent of its members (Blair, 2013). In that sense, the real entity view explicitly recognizes the corporation as an agent in its own right.[3]

A final reason to focus on firms is that the different views of the ontology of the corporation play an important role in contesting what the purpose of firms in society should be. Proponents of the aggregate view typically put it in service of advocating that firms should be managed to serve the financial interests of their shareholders (e.g., Jensen, 2002), while proponents of the real-entity view of the firm tend to argue that firms should be managed to serve the interests of a broader set of stakeholders (e.g., Blair & Stout, 1999). The outcome of the debate between these camps has direct implications for how firms function in practice. On the aggregate-cum-shareholder-wealth-maximization view, managers should run firms to maximize profits, while on the real-entity-cum-stakeholder view, these same managers should balance the interests of all their stakeholders. In fact, opponents of the shareholder wealth maximization doctrine typically see that doctrine as the main cause for turning the firm into an institution that chases profits while imposing negative externalities on society (e.g., Davis, 2021). Insofar as they are right, therefore, ontological discussions about the nature of the firm as a social institution are integral to how modern society functions.

10.3 The Evolutionary Explanation of Human Behavior and Organization

Before turning to an evolutionary analysis of the social ontology of the modern firm, we need to put in place the building blocks for this exercise. My approach to analyzing the modern corporation will rest on two pillars: the concepts of "ontological continuity" and "ontological commonality".[4] Ontological continuity is the idea that, from an evolutionary perspective, we need to understand the world around us as the result of an unbroken chain of causal events.[5] This means that we need to understand our modern institutions, including the corporate form, as phenomena that are both enabled and constrained by the evolutionary history of our species. Ontological commonality is the idea that all evolution, be it biological or cultural, proceeds on the same causal logic. This means that the evolution of cultural phenomena like the modern firm not only can but needs to be understood in similar terms as the evolution of organisms.[6]

Let me first unpack ontological commonality. If all evolution follows the same causal logic, then what is this logic? Answering this question calls for a generalization of evolution that transcends biological and cultural evolution. A particularly insightful attempt at such a generalization is Plotkin's (1994) version of evolutionary epistemology, which conceptualizes evolution as the general process that leads to "knowledge", where knowledge can be understood as encoded information about "what works" in a particular environment – or what evolutionary theory has traditionally referred to as "adaptive fit" (Plotkin, 1994).[7] On this view, evolutionary theory ultimately is a theory of the causal logic that explains how complex adaptive systems maintain (or fail to maintain) their functional integrity over time. Specifying this causal logic requires a combination of two things (Stoelhorst, 2008): the "variation-selection-retention" (V-S-R) algorithm and a distinction between a system's "behavior" and its "codex".

The V-S-R algorithm summarizes how a complex adaptive system is able to maintain adaptive fit in a changing environment. This is the result of cycles of trying out variations in behavior, selecting those behaviors that work, and retaining the ability to reproduce those behaviors. The algorithm can operate when the system is able to encode the instructions for the behaviors that work in the system's codex, so that these behaviors can be reliably reenacted. In essence, then, evolution is the process of feeding back the relative success of the system's behavior (i.e., the success or lack thereof of the system in interacting with its environment) into the system's codex (i.e., the instructions for the system's behavior). For instance, a species evolves because of variations in its phenotype, selection of the phenotypic variations that support behaviors that work in the local environment, and retention of the variations that work in the species' genotype. Similarly, individual organisms can learn because of trying out different ways of interacting with the environment, selecting those ways of interacting with the environment that work, and retaining those ways of interacting with the environment in their neural networks. What differs in these two cases is the timescale on which the feedback cycles of variation–selection–retention operate: in the case of species, each generation constitutes one cycle, while in the case of individual learning, each interaction with the environment may constitute a cycle.

Next, consider ontological continuity. Where do all the various complex adaptive systems in our world come from? How can all these systems, ranging from the simplest single-celled organisms to more complex life forms, and from more complex life forms to complex socio-cultural arrangements like the modern firm, have resulted from one and the same causal chain of events? This brings us to the literature on gene–culture co-evolution (Richerson & Boyd, 2005), multi-level selection theory (Sober & Wilson, 1998), and the theory of major transitions in evolution (Maynard Smith &

Szathmáry, 1995). Together, these theories explain the principles by which evolution begat systems of ever-increasing complexity. Below, I will draw on these three theories to better understand the nature of the modern corporation. However, the central role in my analysis will be played by the theory of major transitions.

The theory of major transitions in evolution (Maynard Smith & Szathmáry, 1995) explains transitions in the complexity of living organisms in terms of "changes in the way information is stored and transmitted" (Szathmáry & Maynard Smith, 1995: 227). Major transitions in evolution, such as the emergence of multi-cellular life forms from single-celled life forms, share three common features (Maynard Smith & Szathmáry, 1995; Szathmáry & Maynard Smith, 1995): first, a "division of labor" (Smith, 1776) – that is, an increase in efficiency due to task specialization; second, a new "language" – that is, a new way of storing or transmitting information; and third a "transition in individuality" (West et al., 2015) – that is, a process resulting in a new entity that acts as a maximizing agent in its own right.[8]

In what follows, I will apply the notions of ontological continuity and ontological commonality to an analysis of the modern firm: I will explain the modern firm as resulting from an unbroken causal chain of events driven by a process of gene–culture co-evolution that has led to a major transition in cultural evolution. This major transition has resulted in a new level of complexity in human social organization in the guise of the corporate form, which creates a legal person that can act as a maximizing agent in its own right. For the corporate form, the "codex" containing the instructions for "what works" is external to the unit of selection (i.e., external to the individual firms that become legal persons) and takes the form of a legal specification of social "roles" and behavioral "norms" related to these roles. One of the implications of this is that modern firms are not reducible to the characteristics of their individual members. The effect of the legal codex is that modern firms have cultural group-level traits that are independent of their individual members. These traits are relatively stable over time but are also continuously contested in an ongoing process of coalitional competition in which subgroups attempt to impose their norms on the group as a whole.

10.4 The Corporation as a Major Cultural Transition

In their analysis of evolutionary transitions in biology, West et al. (2015) note that such transitions involve two steps: "cooperative group formation" and "group transformation". Cooperative group formation is the process by which individual-level cooperative traits emerge, which allows a cooperative group to form. Group transformation is the process that leads to the actual transition to a higher level of "individuality". This transition involves a division of labor among the members of the group, interdependence, coordination, and the suppression of within-group conflict to the point that the group can be considered as acting as one agent and becomes "a fitness-maximizing agent . . . in its own right" (West et al., 2015: 10117). I will follow this distinction between cooperative group formation and group transformation, in the sense that I will first consider cooperative group formation in humans, and will then turn to the cultural innovation of the legal form of the corporation as an instance of group transformation.[9]

10.4.1 Cooperative Group Formation: Explaining Human Cooperation

It may be clear that modern firms are cooperative groups. From an evolutionary perspective, this immediately raises a fundamental question: how to explain that humans are able to maintain such groups? The evolution of cooperation in nature in general, and the evolution of human cooperation in particular, was a central puzzle for both the biological and social sciences in the second half of the 20th century. It would seem that the evolutionary V-S-R algorithm can only reward behaviors that are in the interest of an individual organism: what is retained are behaviors that increase an organism's relative reproductive fitness. Therefore, biological evolution seems to be driven by individual-level competition among organisms, where the winners are those individuals that outcompete other individuals in terms of the number of viable offspring that they produce. How can such an individual-level competitive process possibly give rise to cooperative behaviors, defined as behaviors that benefit the reproductive success of other individuals? Surely, individuals who display such cooperative behaviors would be outcompeted by selfish individuals who do not? These selfish individuals would benefit from the behaviors of their cooperative counterparts without incurring the costs of displaying such behaviors themselves, and would therefore have the advantage in the competition for leaving more offspring. So, even if cooperative individuals would emerge as the result of variation, selection would favor the selfish individuals, and over time, the genetic information that codes for the behaviors of the selfish as opposed to the cooperative individuals would be retained.

While this is indeed the baseline logic of evolution, cooperative behaviors are nevertheless relatively widespread in nature. The resulting puzzle has been solved by an array of work spanning biology, the social sciences, and game theory, which has resulted in the recognition of five mechanisms by which cooperation can emerge out of a competitive process (Nowak, 2006). These mechanisms are "kin selection", "direct reciprocity", "indirect reciprocity", "spatial selection", and "multi-level selection".[10] Kin selection (or "inclusive fitness") is based on the fact that natural selection can favor cooperation when cooperative behavior benefits individuals who are genetically related. Direct reciprocity is based on the fact that natural selection can favor cooperation when individuals play "tit-for-tat" in repeated interactions (i.e., cooperative behaviors are rewarded with cooperative behaviors, and selfish behaviors are punished with uncooperative behaviors). Indirect reciprocity extends this mechanism beyond dyads by using reputation: someone known to be helpful to others is more likely to be met with cooperative behaviors, and someone known to be selfish with uncooperative behaviors.

These three mechanisms very likely played a role in the evolution of human cooperation, but cannot fully explain our cooperative instincts: humans also tend to display cooperative behaviors in one-shot interactions with unrelated strangers in situations without a reputation mechanism (Richerson & Boyd, 2005; Henrich & Muthukrishna, 2021). Hence, to fully explain human cooperation, we also need to invoke spatial selection and multi-level selection. Spatial selection is based on cooperators forming network clusters that increase the relative frequency of their interactions with other cooperators, which allows them to outcompete selfish individuals. The role of spatial selection in human cooperation is understudied, but I will have something to say about it in the conclusion of this chapter. Multi-level selection is based on the recognition that competition may take place not only between individuals but also between groups. In contrast to spatial selection, the role of multi-level selection in the evolution of human cooperation has received a lot of attention, in particular in relation to the central role of culture in human affairs (Henrich & Muthukrishna, 2021).

The central idea in multi-level selection (MLS) theory is aptly summarized as "Selfishness beats altruism within groups. Altruistic groups beat selfish groups. Everything else is commentary" (Wilson & Wilson, 2007: 345). MLS theory makes a distinction between within-group competition, in which individuals displaying selfish behaviors have the competitive advantage, and between-group competition, in which groups displaying cooperative behaviors have the competitive advantage. The crux of the theory is to recognize that "[t]otal evolutionary change in a population can be regarded as a final vector made up of two component vectors, within- and between-group selection, that often point in different directions" (Wilson & Wilson, 2007: 330). That MLS must have played an important role in the evolution of human cooperation becomes clear when we include cultural evolution into the analysis, which brings us to "dual inheritance" or "gene-cultural co-evolution" theory (Richerson & Boyd, 2005).

Gene–cultural co-evolution theory recognizes two inheritance systems, or more accurately, two ways of transmitting information about what works: genetic transmission and cultural transmission. Genetic transmission takes place when organisms reproduce, while cultural transmission takes place when organisms learn from others. Cultural transmission substantially increases the likelihood of making between-group competition an evolutionary force because it both decreases behavioral variation within groups and increases behavioral variation between groups (Richerson & Boyd, 2005). These conditions are necessary for between-group selection to "overcome" the selective advantage of selfish behaviors in within-group competition. Once the human species evolved a capacity for culture (i.e., a capacity to learn from others), groups with more cooperative cultures could outcompete groups with less cooperative cultures, which changed the vector of selection pressures on individual behaviors in a way that allowed individuals to evolve cooperative traits.

Tomasello (2016) has proposed a plausible scenario that details how our culturally enabled way of cooperating may have evolved.[11] Building on insights into how human behavior differs from the behavior of our closest primate cousins (Tomasello & Vaish, 2013), this scenario involves three steps in the development of human social cognition. The first step is the evolution of "joint attention". Joint attention is an obvious precondition for cooperation.[12] This step was followed by two more steps of cognitive evolution that resulted in what we tend to refer to as human morality: first, the evolution of a psychology that allowed for "mutualistic collaboration and prosocially motivated interactions with specific other individuals", and second, the evolution of "the more abstract, agent-neutral, norm based morality of individuals who live in more large-scale cultural worlds full of impersonal and mutually known conventions, norms, and institutions" (Tomasello & Vaish, 2013: 232). What is especially interesting for our understanding of the aspects of human psychology that underwrite our unique ability to cooperate is that Tomasello (2016) links the second step (the ability to cooperate with specific others) to the ability to recognize *roles* (which support a division of labor), and the ability to "honor" these roles (i.e., the willingness to display the behaviors required by the roles). Similarly, he links the third step (the ability to cooperate in larger groups) to the ability to recognize *norms* and the ability to enforce them (i.e., the ability to both discipline oneself and others to behave in accordance with the norms).

It may be clear that the evolution of a role-based and norm-based psychology was a boon to cooperation. In essence, what the evolution of human social and moral psychology achieved is to suppress selfish behavior in favor of behavior that benefits the group: human morality evolved to support living in groups (Rai & Fiske, 2011). Reflecting this, our so-called "moral sentiments" (Smith, 2010[1759]) have been described as "[m]otives such as altruism, reciprocity, intrinsic pleasure in helping others, aversion to inequity, ethical commitments, and other motives that induce people to help others more than is consistent with maximizing their own wealth or material payoff" (Bowles, 2016: 45). Human moral psychology allows cooperative groups to form by suppressing within-group conflict. At the same time, our ability to use roles and norms to form cooperative groups also helped create the conditions for between-group competition. In line with this, one of the downsides of human morality is that we combine an instinct to cooperate with in-group members with an instinct to compete (or worse) with out-group members (Greene, 2014). That human morality takes this particular form is consistent with a role for between-group competition in the evolution of human cooperation.

Where does this leave us with respect to understanding the corporation as a major transition? We have reached an understanding of how humans became able to cooperate at all, which is the necessary first step for understanding a major transition. In the case of humans, what West et al. (2015) refer to as "cooperative group formation" took place as the result of a multi-level selection process in which culture (information transmitted through learning from others) helped create the conditions for between-group competition to become an important force. In this between-group competition, groups with more cooperative cultures could outcompete groups with less cooperative cultures. Such cooperative cultures were underwritten by group-specific social roles and behavioral norms. Cooperative norms both changed within-group selection pressures in favor of cooperative behaviors and gave the groups that evolved them an advantage in competition with other less cooperative groups. Maintaining these cooperative norms was enabled by the evolution of the role-based and norm-based psychology that we traditionally refer to as human morality.

We can already discern most aspects of a major transition in this first step: a division of labor, the emergence of a new "language" (a new way of transmitting information), and the suppression of within-group conflict. The division of labor takes the form of individuals assuming specific roles, and the new way of transmitting information is spoken language. A crucial part of this language is the moral language that suppresses selfish behaviors in favor of behaviors that benefit the group. What we do not yet see is the completion of the second step in a major transition: a full-fledged transition in individuality. Below, I will argue that this second step has begun to take shape with the evolution of the modern firm.

10.4.2 *Group Transformation: Explaining the Modern Firm*

Following West et al.'s (2015) distinction, group transformation is the process by which a cooperative group becomes an "individual" in its own right. This second phase of a major transition completes the actual transition: it results in a new and more complex entity that can be understood as an agent that interacts with its environment in ways that increase its own fitness. This is what happened once firms could be organized as corporations. The corporation is the archetypal "legal person". It has its roots in Roman law, but it emerged in modern form in the 19th century (Blair, 2013). What makes the corporation interesting from the point of view of the literature on major transitions is that "[w]hen a corporation is formed, the law immediately recognizes the existence of a new legal entity that is separate from the organizers and investors, an entity that can carry out certain business activities as a 'person'" (Blair, 2013: 786). Corporate law quite literally creates a new type of agent.

While it may at first glance seem a leap to move from a discussion of the emergence of human cooperation in the early days of human evolution to the emergence of the corporate form in the 19th century, all that we need to connect the two is the same evolutionary logic that we already employed: a multi-level selection logic reinforced by the cultural transmission of information. An important consequence of adding the cultural transmission of information into the mix of evolutionary mechanisms is not just that it helped create the conditions for between-group competition to play an important role in changing the vector of selection pressures on individual behavior in favor of cooperation. Cultural transmission also created the conditions for what Muthukrishna (2023) refers to as "cultural-group selection". The emphasis in this concept is on the hyphen: once culture starts playing a role, selection no longer only acts on individual-level traits (like selfish or cooperative behaviors), but also on group-level traits. Different groups have different culturally evolved group-level traits that are independent of the traits of their individual members, such as a particular way of organizing the group. In a multi-level selection context, the competition between groups also needs to be understood as a competition between these group-level traits.[13]

Once we take this view, the only other thing that we need in order to understand the nature and emergence of the corporate form is the recognition of how written language was used to institutionalize group-level traits that "worked". While written language originally evolved for bookkeeping purposes, it took on an additional role when it was used to codify behavioral norms, as it did in religious and legal texts. Written language offered another way to encode and transmit information about what worked than transmitting information directly from individual to individual (or, more specifically, directly from brain to brain) by spoken language. Moreover, information encoded in text had some obvious advantages in terms of the fidelity of transmitting the information and in terms of how easily information could be transmitted across space and time.

Written information in the form of legal texts can be understood as one of the "cranes" in cultural evolution that led to the corporate form.

Dennett (1996) coined the term crane to refer to innovations that enable the evolutionary process to increase organizational complexity. The idea is that biological evolution has occasionally "hit" on inventions that would subsequently allow the complexity of life to increase. Generalizing this idea to cultural evolution helps understand how an accumulation of cultural innovations ultimately led to the corporate form and the introduction of a higher-level agent, the hallmark of what West et al. (2015) refer to as the group transition phase of a major transition.

Although Blair (2013) does not use evolutionary language, she discusses the evolution of the corporate form in functional terms that are directly in line with the V-S-R logic of evolutionary theory. Her analysis is based on pointing out how the corporate form emerged to fulfil four functions that helped entrepreneurs and investors to more efficiently organize business activities. In other words, her analysis is couched in terms of a process by which, over time, the corporate form emerged as a way of organizing business activity that "worked". The four functions are (1) providing continuity and a clear line of succession in property and contract, (2) providing an "identifiable persona" to serve as a central actor in carrying out the business activity, (3) providing a mechanism for separating pools of assets belonging to the corporation from those belonging to the individuals participating in the enterprise, and (4) providing a framework for self-governance of certain business or commercial activity (Blair, 2013: 785). Blair's analysis is one in which cultural evolution, over time, "hit" on a number of cultural cranes that enabled a legal form, the corporation, that fulfilled these four functions. Central to these cultural cranes is the concept of "legal personhood" (also referred to as "legal personality" or "entity status").

Incorporation creates a legal "person" that can, in principle, have perpetual life. This perpetual life is achieved by perpetual succession: the legal person can continue to exist even as individual people move in and out of the firm. Importantly, in order to achieve this, once incorporated, the firm as a legal person can own assets and can enter into contracts. To enable this, it can sue and be sued. That it is the firm itself, as opposed to its individual members, that can sue and be sued, is underwritten by two more cultural cranes: "limited liability" and "capital lock-in". Capital lock-in means that shareholders cannot extract capital from the corporation: once invested equity is owned by the firm itself (Blair, 2003). Limited liability means that shareholders are only liable for the amount of capital that they have invested in the firm. Together, these two mechanisms are referred to as "asset partitioning" (Hansmann & Kraakman, 2000), and they help create a clear boundary between the incorporated firm as a legal person, on the one hand, and its various stakeholders, on the other hand. Finally, once a separate entity is thus created, it requires a governance structure that specifies who can act on behalf of the legal person. Here, the cultural crane is the concept of an independent board of directors (i.e., an organ that is independent of the firm's shareholders, managers, employees, and other stakeholders) and that owes a fiduciary duty to the interests of the corporation itself (Blair & Stout, 1999). An additional cultural crane that reinforces the individuality of the corporation is the so-called "business judgment rule", which holds that courts will defer to the corporation's board on matters of internal governance unless it is clear that the members of the board did not act in good faith and with their fiduciary duty to the corporation in mind.

Over time, these various cultural cranes enabled the specific configuration of legal rules that define the modern corporation. This particular configuration evolved out of a process of variation, selection, and retention: different ways of organizing business activities were tried (variation), the ways of organizing business activities that worked survived (selection), and these ways of organizing business activities were encoded in legal rules (retention). The corporate form did not, like Pallas Athena, emerge fully armored from the brow of Zeus but was tinkered together over time as the result of a process of cultural evolution. In fact, even today, there is not one corporate form, as different jurisdictions tend to have somewhat different rules, so some variation persists. But there is clearly a family resemblance across jurisdictions in terms of specific cultural group traits that define a corporation. And it is these particular group traits that explain why the corporate form replaced other legal forms, such as the partnership, as the dominant way of organizing business activity. Simply put, since the mid-19th century, the corporate form is what "worked", in particular because it allowed the organization of business activities that (1) called for a long-term orientation; (2) required large-scale capital investments; and/or (3) relied in substantial part on intangible assets (such as technological knowledge, reputation, or organizational culture).

Among the various cultural group traits that corporate law encodes, legal personhood (the corporation's status as a legal person in its own right) has been central to the success of the corporation. Legal personhood is also the trait that qualifies the corporation as a major transition in evolution. Not only is the corporation a legal person it its own right, but its internal governance is, by law, organized in such a way that it is managed by an organ (the board) that is independent of the corporation's stakeholders and that, again by law, is held to further the interests of the corporation itself (Blair & Stout, 1999). It is in this sense that the corporation qualifies as a fitness-maximizing agent in its own right, the main criterion for a major transition according to West et al. (2015). That corporations tend to behave in their own interest, as opposed to the interests of their individual stakeholders, or the interests of society as a whole, is also the main reason why their role in our world is the topic of heated debate.

10.5 From the Cultural Evolution of the Modern Firm to Social Ontology

How does the evolutionary analysis above inform social ontology? I will first translate this analysis into an answer to the question "what is a firm", and will then generalize that answer to what it implies about the nature of human social organization more generally. In doing so, I will make a distinction between firms in general and incorporated firms. I will first answer the question "what is a firm" for firms in general, both from the point of view of ontological continuity and from the point of view of ontological commonality. I will then do the same for corporations. I will also reflect on the nature of the culture-driven evolutionary process that is a source of the continuous change of social reality.

10.5.1 The Ontology of the Firm

From the point of view of ontological continuity, a firm is "merely" an example of the human ability to cooperate, as are so many other human institutions, including families, sports teams, political parties, and armies. In other words, at heart, the firm is simply a cooperative group. The human ability to form cooperative groups evolved as the result of a process of simultaneous within-group and between-group competition, which gave rise to a moral psychology that revolves around social roles and behavioral norms. This moral psychology co-evolved with spoken language and suppresses within-group competition. Firms capitalize on the evolutionary origins of human psychology in two ways: they appeal to our sensitivity to roles and norms to enable within-group cooperation (see Tomasello, 2016), but they also appeal to our sensitivity to group identities (Greene, 2014) to allow us to enact a peaceful form of between-group competition. Given the evolutionary origins of human nature, it should not come as a surprise that the institution at the heart of modern-day economies (i.e., at the heart of our contemporary survival strategy) is an institution that creates welfare out of simultaneous within-group cooperation and between-group competition. The legal rules that evolved to make corporations possible were both enabled by the particular nature of human psychology and reinforced that psychology.

From the point of view of ontological commonality, each individual firm is a complex adaptive system. Like all complex adaptive systems, it has a "design" that specifies its component parts, as well as how these parts interact with each other. In the case of the firm, its component parts are individuals who take on particular roles. Behavioral norms related to these roles specify how individuals should interact. The various roles implement a division of labor, and the behavioral norms allow the firm as a whole to display so-called "routines" (Nelson & Winter, 1982), which can be understood as culturally transmitted collective behavior patterns that can reliably be triggered in response to specific environmental cues. It is on the basis of these collective action patterns (i.e., on the basis of the particular ways in which individual firms interact with their environment) that firms compete with each other. In other words, firms compete on the basis of culturally evolved group-level traits.

What do these insights imply for how we are to understand the nature of social reality more generally? For as far as firms are representative of how human groups function (and I believe that unincorporated firms are), their nature implies that we can only understand social reality if we explicitly recognize how the individual and collective levels of analysis interact. Social reality consists of individuals, but individuals can and do form groups that have traits that are not reducible to the traits of the individual members of those groups. The collective behavior of a social group, such as a firm, is not simply an aggregation of the behaviors that its members would display outside of the group. A social group is not a mob. The difference between a mob and a social group like a firm (or a family, sports team, political party, army, etc.) is that the firm has a social structure: members of the firm adapt their behaviors to the roles and norms of the group.

In general terms, then, we should understand social reality in terms of roles that enable the division of labor and norms that enable predictable collective action patterns. Roles and norms are the linking pin between the individual and collective levels of analysis.[14] Collective action patterns that work (i.e., collective action patterns that lead to successful interactions with the environment) become encoded in roles and norms. These roles and norms are group-level traits. They are culturally transmitted (i.e., by learning from others) and define the social structure of the group. In other words, roles and norms are the "codex" of a social group. They provide stability to the group's collective action patterns, so that successful ways of interacting with the environment can be reproduced. They are also what evolves over time as successful variations in the group's collective action patterns lead to adaptations.[15]

10.5.2 *The Ontology of the Corporation*

An evolutionary analysis of the corporate form leads to a number of additional insights about the nature of social reality. Whereas firms in general can be understood as cooperative groups, modern firms that adopt the corporate form need to be understood as agents in their own right. In terms of ontological continuity, the cultural innovation of legal personhood signals that human group formation has entered the second phase of a major transition, the phase in which cooperative groups are transformed into entities that maximize their own interest, as opposed to merely aggregating the individual interests of their members. That corporations can function this way is underwritten by the same moral psychology that enables firms (or cooperative groups in general) to suppress internal conflict, namely our instinct to behave in line with roles and group norms.

In terms of ontological commonality, the transformation of the firm into a corporation with legal personhood status requires the legal scaffolding of corporate law. What corporate law does is provide a codex that is external to the firm itself. This codex does not specify all the roles and norms that an individual firm needs in order to develop the specific collective behaviors that will allow it to successfully compete in, say, making steel or selling hamburgers. But it does offer a legal template that specifies a number of essential roles (the board, shareholders, employees) and the way that the individuals in these different roles can interact. In addition, and crucially, it also grants firms that adopt this template the legal rights that allow them to function as agents in their own right.[16]

What do these additional insights into the nature of the corporation imply for our understanding of the nature of social reality more generally? The nature of the corporation illustrates two important points about social ontology. The first is that important parts of social reality are "institutionalized": the roles and rules that define different types of cooperative groups (families, sports teams, political parties, armies, and different types of firms) are encoded in written legal texts that are external to the groups in question. Moreover, this legal scaffolding for the design of cooperative groups is backed up by force as a result of the monopoly on violence of the state. This provides a backstop to be able to suppress behaviors that deviate from the interests of the legally defined group.

The second point that the nature of the corporation illustrates about social ontology is that the scaffolding of the law makes it possible to transform cooperative groups into agents in their own right. It is the law that enables the completion of the human major transition. This does not just apply to incorporated firms, but also to other kinds of legal persons, including the state. Corporations and states need to be understood as agents that will pursue their own interests. A crucial corollary to this conclusion is that social reality will in large part be determined by how artificial agents such as corporations and states define their interests. To understand how they do, we need to zoom out and consider the multi-level dynamics of cultural evolution.

10.5.3 *The Nature of Cultural Evolution*

In addition to considering the nature of the firm in general and the corporation in particular, it is also instructive to consider the nature of the evolutionary process that brought them about and, importantly, continues to change them. The corporation emerged as the most important economic institution of our time as a result of a process of competition between different legal forms. Moreover, these legal forms evolved over time, both as a direct result of this competitive process, in which favorable variations would be selected (variations that worked in the specific local and historical context) and as a result of the emergence and assimilation of new cultural "cranes", such as capital lock-in, limited liability, or the concept of an independent board. These same processes (competition between different legal forms and the emergence and assimilation of cultural innovations) are going on today. As the product of an evolutionary process, social reality is never static.

In line with our earlier analysis, we can and should understand social reality as the ever-changing product of a culture-enabled multi-level selection process. Considering the contested nature of the corporation helps bring the nature of that process into focus. What is contested is not so much the existence of the corporation as such, but its "purpose": what a corporation's role in society should be. Remember that the board has a fiduciary duty to the corporation itself. What this means is that the role of the board is to further the interests of the corporation. But what are these interests? In essence, what is contested is what the board's objective function for running the firm should be.

The form that this contestation takes is best understood as a form of "coalitional competition". What I mean by this is a process in which subgroups try to "capture" the corporation by imposing behavioral norms that are in their coalitional interest. Human social groups in general, and the corporate form in particular, are generally effective in suppressing within-group competition. In the case of the corporation, this is in part because of a reasonably clear definition of the roles of the board, shareholders, employees, and other stakeholders. What is contested are not so much these roles. In particular, the primacy of the board is rarely contested. But subgroups within the corporation do compete to have their coalitional interests reflected in the objective function of the firm.

Given the fact that the template for the corporation takes the form of an external legal codex, most of this contestation takes the form of coalitional competition to change legal rules, or at least the interpretation of these rules. Since the 1980s, a coalition that has put the interests of shareholders central has been quite successful in imposing shareholder wealth maximization as the "proper" objective function on corporations, especially in the United States (Jensen, 2002). But despite the declaration of "the end of history in corporate law" in 2001 (Hansmann & Kraakman, 2001), a declaration that claimed that shareholder primacy now effectively reigned supreme in defining the purpose of the corporation, developments since then have shown that other coalitions which put more weight on the interests of employees or society as a whole are able to contest this interpretation of corporate law.[17]

What the example of the contested nature of the corporation tells us about the nature of social reality more generally is that we not only need to understand this reality as the product of a multi-level selection process in which there is competition within groups and competition between groups, a combination that can support the formation and maintenance of cooperative groups. We also need to understand the role of competition between coalitions: subgroups (such as activist shareholders) that may compete to capture a cooperative group by imposing behavioral norms that are in their coalitional interest. Success in this coalitional competition is what underwrites power: the ability to define the behavioral norms for the group as a whole.

10.6 Conclusion

In this chapter, I have explored how an evolutionary explanation of human cooperation in general, and human cooperation in modern firms that have adopted the corporate form in particular, can inform social ontology. Insofar as social ontology is an exercise in carving social nature at its joints, the evolutionary explanation of human cooperation suggests that we should recognize the difference between three types of groups: a group of people, people in a group, or a group that is a maximizing agent in its own right. A mob is a group of people, a firm is people in a group, and a corporation is an agent in its own right. What distinguishes a firm from a mob is that the firm has a social structure that can be understood in terms of culturally transmitted roles and norms that, together, underwrite predictable group behaviors. What distinguishes a corporation from a firm is that the corporation has a life and purpose that is independent from its members, who merely enact the roles and norms that an external legal codex specifies. However, the purpose of the corporation (what the role of the corporation in society is, and how a firm should be managed to fulfil this role), as any social norm, is up for contestation and will evolve as the result of a continuous process of coalitional competition in which subgroups compete to impose their own norms on the group as a whole. Success in this process of coalitional competition is what leads to power.

In addition to shedding light on the nature of the modern firm and social ontology more generally, my analysis makes two contributions to the literature on cultural evolution. The first is to show how the theory of the major transitions in evolution, originally developed to account for transitions in biological evolution, can be extended and generalized to cultural evolution. Like biological evolution, cultural evolution has similarly hit on "cranes" that have opened up new evolutionary avenues that led to more complex forms of organization with their own individuality, such as the corporation. Among these cranes are written language, legal texts, and a number of legal innovations (such as capital lock-in, limited liability, and a board with a fiduciary duty to the corporation itself) that underwrite the corporate form as a legal person and a maximizing agent in its own right. The second contribution to the literature on cultural evolution is to show how taking ontological continuity and ontological commonality seriously leads to a view of cultural evolution that overcomes the critiques that cultural evolution theory tends to take a methodologically individualistic approach (Valković, 2024) and that it negates the role of power (Fracchia & Lewontin, 1999).[18] The combination of gene–culture co-evolution theory, multi-level selection theory, and the theory of major transitions that I have put to work leads to an explicit recognition of the importance of understanding the competition between groups in terms of cultural group-level traits that are independent of their members, as well as the importance of understanding how the behavioral norms of groups are contested to the point that groups may be captured by subgroup coalitions that succeed in imposing their norms on the group as a whole.

Acknowledgments

With thanks for the helpful comments of Bart Geurts, Thomas A.C. Reydon, Marc Slors, and Martina Valković. The usual disclaimer applies.

Notes

1 For a related but more general analysis of the relevance of evolutionary theory for understanding institutions, see the chapter by Hindriks in this volume.
2 On the topic of power, also see the chapters of Baraghith and Jordan, Burman and Pelgrom, Chellappoo, and Lohse and Valković in this volume.
3 The third of the competing views, the concession view, sees the corporation as a concession that is granted by a central authority (the King or the State). This is how the precursors of the modern corporation, such as the Dutch VOC (East Indies Company), emerged. What is central to the concession view is the idea that the special privileges that are bestowed on the corporation, such as, historically, a monopoly, or later, limited liability, serve (or should serve) some public benefit.
4 The label "ontological continuity" is adopted from the work of Ulrich Witt (2004) and the label "ontological commonality" from the work of Geoffrey Hodgson (2004) – see Stoelhorst (2008).
5 This aligns with the "one world" idea in Searle's approach to social ontology (see Searle, 1998).
6 In essence, the idea of ontological commonality is that studying cultural evolution is not merely an exercise in metaphor or analogy based on transferring (elements of) our understanding of biological evolution to cultural evolution. It is

rather about deriving a general understanding of the causal logic underlying all evolutionary processes by generalizing *across* biological evolution and cultural evolution. If all evolution involves one and the same generic causal logic, then the key to any explanatory project related to (biological or cultural) evolution is to understand and apply that logic.

7 The term evolutionary epistemology is often primarily associated with an approach that applies an evolutionary logic to illuminating the nature of individual (Campbell, 1960) or scientific knowledge (Popper, 1972). Plotkin's (1994) version goes further and essentially sees evolution as a process of knowledge acquisition. It is this feature of his more radical version of evolutionary epistemology that makes it particularly well-suited as the starting point for an explanatory project that takes both ontological continuity and ontological similarity seriously. On Plotkin's view, evolution=learning and knowledge=adaptive fit. Applied to firms and institutions more generally, this means that their features can be understood as products of an evolutionary process that has favored "what works" in a particular local environment. For discussions of the relationship between Plotkin's approach and the approach of Campbell and Popper, see Bradie (1986) and Gontier & Bradie (2021).

8 The West et al. (2015) understanding of a major transition in terms of the emergence of a new entity that acts as a maximizing agent in its own right is somewhat broader than the original definition in Maynard Smith & Szathmáry (1995), which emphasizes how entities that were previously capable of replicating independently are only able to replicate as part of the larger unit that results from the transition. In the terminology of Godfrey-Smith (2009), the West et al. definition emphasizes "Darwinization" (i.e., a (higher-level) entity becoming a unit of selection in its own right), whereas the Maynard Smith & Szathmáry definition emphasizes "de-Darwinization" (i.e., a (lower-level) entity losing its status as a unit of selection in its own right). As we will see below, in the case of the firm there is Darwinization (of the firm) without de-Darwinization (of its members).

9 There is a close relationship between these two phases and the distinction between the two interpretations of multi-level selection: MLS1 and MLS2 (Okasha, 2005). The concept of multi-level selection will be explained below. In MLS1, the focus is on the relative frequency of different entities within a collective, whereas in MLS2, the focus is on the relative frequency of the different types of collectives themselves. This distinction is important for transcending the traditional individualistic view of cultural evolution theory.

10 The labels used here are the updated labels from Rand & Nowak (2013). Nowak (2006) uses the labels "network selection" for spatial selection and "group selection" for multi-level selection.

11 For a more detailed discussion of Tomasello's work, see the chapter by Slors in this volume.

12 For a more general discussion of the preconditions for human cooperation, see the chapter by Geurts in this volume.

13 Note that it is at this point (i.e., when the role of culturally evolved group-level traits enters the analysis) that cultural evolution theory transcends its original methodological individualism roots.

14 For a similar perspective, see the chapter by Hindriks in this volume.

15 The sources of variations in a firm's collective action patterns can be more or less random or directed. They include replacing a particular individual who

performs a particular role with another individual, who may not enact the role in exactly the same way, but also more conscious attempts to change the collective action patterns, such as a reorganization.

16 For a detailed analysis of exactly how far this has gone in the United States, where, in the (in)famous "Citizens United" and "Hobby Lobby" cases, the Supreme Court has granted corporations far-reaching rights that include the right to support political campaigns, see Winkler (2018).

17 One consequence of this contestation was the highly publicized 2019 statement of the influential U.S. corporate lobby group "The Business Roundtable" (an organization that had, since 1997, consistently advocated shareholder primacy), which redefined the purpose of the firm in stakeholder terms.

18 On the role of power in cultural evolution, also see the chapters of Baraghith & Jordan, Burman & Pelgrom, Chellappoo, and Lohse & Valković in this volume.

References

Blair, M.M. (2003): Locking in capital: What corporate law achieved for business organizers in the nineteenth century, *UCLA Law Review* 51: 387–456.

Blair, M.M. (2013): Corporate personhood and the corporate persona, *University of Illinois Law Review* 3: 785–820.

Blair, M.M. & Stout, L.A. (1999): A team production theory of corporate law, *Virginia Law Review* 85: 247–303.

Bowles, S. (2016): *The Moral Economy: Why Good Incentives Are No Substitute for Good Citizens*, New Haven: Yale University Press.

Bradie, M. (1986): Assessing evolutionary epistemology, *Biology and Philosophy* 1: 401–459.

Campbell, D.T. (1960): Blind variation and selective retention in creative thought as in other knowledge processes, *Psychological Review* 67: 380–400.

Davis, G.F. (2021): Corporate purpose needs democracy, *Journal of Management Studies* 58: 902–913.

Dennett, D.C. (1996): *Darwin's Dangerous Idea: Evolution and the Meanings of Life*, New York: Simon and Schuster.

Fracchia, J. & Lewontin, R.C. (1999): Does culture evolve?, *History and Theory* 38: 52–78.

Godfrey-Smith, P. (2009): *Darwinian Populations and Natural Selection*, New York: Oxford University Press.

Gontier, N. & Bradie, M. (2021): Evolutionary epistemology: Two research avenues, three schools, and a single and shared agenda, *Journal for General Philosophy of Science* 52: 197–209.

Greene, J. (2014): *Moral Tribes: Emotion, Reason, and the Gap between Us and Them*, New York: Penguin.

Hansmann, H. & Kraakman, R. (2000): The essential role of organizational law, *Yale Law Review* 110: 387.

Hansmann, H. & Kraakman, R. (2001): The end of history for corporate law, *Georgetown Law Journal* 89: 439–468.

Henrich, J. (2004): Cultural group selection, coevolutionary processes and large-scale cooperation, *Journal of Economic Behavior and Organization* 53: 3–35.

Henrich, J. & Muthukrishna, M. (2021): The origins and psychology of human cooperation, *Annual Review of Psychology* 72: 207–240.

Hindriks, F. & Guala, F. (2015): Institutions, rules, and equilibria: A unified theory, *Journal of Institutional Economics* 11: 459–480.

Hodgson, G.M. (2004): Darwinism, causality and the social sciences, *Journal of Economic Methodology* 11: 175–194.
Jensen, M.C. (2002): Value maximization, stakeholder theory, and the corporate objective function, *Business Ethics Quarterly* 12: 235–256.
Jensen, M.C. & Meckling, W.H. (1976): Theory of the firm: Managerial behavior, agency costs and ownership structure, *Journal of Financial Economics* 3: 305–360.
Lawson, T. (2012): Ontology and the study of social reality: Emergence, organisation, community, power, social relations, corporations, artefacts and money, *Cambridge Journal of Economics* 36: 345–385.
Lawson, T. (2015): The nature of the firm and peculiarities of the corporation, *Cambridge Journal of Economics* 39: 1–32.
Maynard Smith, J. & Szathmáry, E. (1995): *The Major Transitions in Evolution*, Oxford: Oxford University Press.
Muthukrishna, M. (2023): *A Theory of Everyone: The New Science of Who We Are, How We Got Here, and Where We're Going*, Cambridge, MA: MIT Press.
Nelson, R.R. & Winter, S.G. (1982): *An Evolutionary Theory of Economic Change*, Cambridge, MA: Harvard University Press.
Nowak, M.A. (2006): Five rules for the evolution of cooperation, *Science* 314: 1560–1563.
Okasha, S. (2005): Multilevel selection and the major transitions in evolution, *Philosophy of Science* 72: 1013–1025.
Plotkin, H.C. (1994): *Darwin Machines and the Nature of Knowledge*, Harmondsworth: Penguin.
Popper, K.R. (1972): *Objective Knowledge: An Evolutionary Approach*, Oxford: Clarendon Press.
Rai, T.S. & Fiske, A.P. (2011): Moral psychology is relationship regulation: Moral motives for unity, hierarchy, equality, and proportionality, *Psychological Review* 118: 57.
Rand, D.G. & Nowak, M.A. (2013): Human cooperation, *Trends in Cognitive Sciences* 17: 413–425.
Richerson, P.J. & Boyd, R. (2005): *Not by Genes Alone: How Culture Transformed Human Evolution*, Chicago: University of Chicago Press.
Runciman, D. (2023): *The Handover: How We Gave Control of Our Lives to Corporations, States and AIs*, New York: Liveright Publishing.
Searle, J.R. (1998): Social ontology and the philosophy of society, *Analyse & Kritik* 20: 143–158.
Searle, J.R. (2005): What is an institution?, *Journal of Institutional Economics* 1: 1–22.
Smith, A. (1776): *An Inquiry into the Nature and Causes of the Wealth of Nations* (Cannan edition), New York: The Modern Library.
Smith, A. (2010[1759]): *The Theory of Moral Sentiments*, Harmondsworth: Penguin.
Sober, E. & Wilson, D.S. (1998): *Unto Others: The Evolution and Psychology of Unselfish Behavior*, Cambridge, MA: Harvard University Press.
Stoelhorst, J.W. (2005): The naturalist view of universal Darwinism: An application to the evolutionary theory of the firm, in: Finch, J. & Orillard, M. (Eds): *Complexity and the Economy*, Cheltenham: Edward Elgar, pp. 127–147.
Stoelhorst, J.W. (2008): The explanatory logic and ontological commitments of generalized Darwinism, *Journal of Economic Methodology* 15: 343–363.
Stoelhorst, J.W. & Richerson, P.J. (2013): A naturalistic theory of economic organization, *Journal of Economic Behavior & Organization* 90: S45–S56.
Szathmáry, E. & Maynard Smith, J. (1995): The major evolutionary transitions, *Nature* 374: 227–232.

Tomasello, M. (2016): *A Natural History of Human Morality*, Cambridge, MA: Harvard University Press.
Tomasello, M. & Vaish, A. (2013): Origins of human cooperation and morality, *Annual Review of Psychology* 64: 231–255.
Valković, M. (2024): Ontological and methodological limitations of certain cultural evolution approaches, *Philosophy of the Social Sciences* 54: 279–301.
West, S.A., Fisher, R.M., Gardner, A. & Kiers, E.T. (2015): Major evolutionary transitions in individuality, *Proceedings of the National Academy of Sciences* 112: 10112–10119.
Wilson, D.S., Van Vugt, M. & O'Gorman, R. (2008): Multilevel selection theory and major evolutionary transitions: Implications for psychological science, *Current Directions in Psychological Science* 17: 6–9.
Wilson, D.S. & Wilson, E.O. (2007): Rethinking the theoretical foundation of sociobiology, *The Quarterly Review of Biology* 82: 327–348.
Winkler, A. (2018): *We the Corporations: How American Businesses Won Their Civil Rights*, New York: Liveright.
Witt, U. (2004): On the proper interpretation of 'evolution' in economics and its implications for production theory, *Journal of Economic Methodology* 11: 125–146.

Part IV

Outlook

11 Bridging Research in Cultural Evolution and Social Ontology

A Perspective on Future Work

Thomas A.C. Reydon and Martina Valković

11.1 Introduction

With this volume, we aim to explore the area where cultural evolution and social ontology intersect topically and to foster the cross-fertilization of approaches in these two fields of research that lead largely separate lives. While the previous chapters have all examined specific aspects of this overarching theme, taken together, they can only constitute a starting point for future work. In this concluding chapter, we want to present a broader perspective on such future work. Note, however, that we are not aiming to provide an exhaustive view of the directions in which research at the intersection of cultural evolution and social ontology could or should go. Our aim is merely to provide suggestions for future work that have the potential to start off new lines of research.

We will do so by first looking back at the chapters of this volume, highlighting some important research questions that ensue from the work presented in these chapters, as well as the connections forged to other fields of research, such as political philosophy, philosophy of the social sciences, anthropology, and others. This will yield a perspective on the kinds of questions that future research at the intersection of cultural evolution and social ontology could (and should) address. We will then add to this perspective by discussing some of the more general connections between (biological) evolution and ontology more generally. We will suggest additional avenues for research that could be taken (even though we also express some skepticism with respect to these avenues). We conclude by addressing some of the methodological challenges that attempts at connecting research in cultural evolution and in social ontology will be faced with, connecting the topic of the volume to the philosophy of interdisciplinarity.

11.2 Research Questions and Connections to Other Fields Raised by This Volume's Chapters

The preceding chapters in this volume raise a variety of research questions that are also relevant for the debates in fields other than cultural evolution

DOI: 10.4324/9781003643845-15

and social ontology. Here we will discuss the most prominent among them. They include questions regarding ontological and methodological individualism and holism, a subject that is highly relevant for the philosophy of social sciences. Other sets of questions raised by the volume concern the concept and role of power in cultural evolution, social ontology, and beyond, which is equally important when it comes to advancing analogous debates in philosophy of the social sciences, sociology, anthropology, and political philosophy. The latter fields will also benefit from the volume's discussion on conceptualizing institutions. Chapters in the volume are also variously informed by rational choice theory, evolutionary game theory, ethics, biology, psychology, philosophy of language, linguistics, and economics, thereby advancing these fields as well.

11.2.1 *Ontology and Methodology: Individualism, Holism, or Pluralism?*

Perhaps the most discernible question that crystallized in the preceding chapters is that of the appropriate approach to either the ontology of sociocultural phenomena, the methodology used to study them, or both.

Already in the second chapter, Lohse and Valković argue that reductive individualism is a prominent tendency in much of cultural evolution research, and go on to present criticism raised against this tendency, including that it creates distorted ontology by neglecting the holistic dimensions of culture and the existence and impact of social structures, and that it provides misleading explanations of social phenomena since it disregards the roles of organized agents and institutional factors. Chellappoo (Chapter 5) elaborates on a version of this criticism, arguing that, while methodological individualism can explain more than we may initially think, it still appears limited in that it does not provide the full picture of the phenomena in question (and that it may omit the more important factors).

Along similar lines, Hindriks (Chapter 4) makes a case for a pluralist approach in the study of cultural evolution, according to which *some* social phenomena can adequately be explained in terms of individual agents and actions. A methodological pluralist will then explain some social phenomena, or some of their aspects, in individualist terms, and others in more complex terms, that is, in terms of social facts and events. Cultural evolutionists who embrace ontological individualism, Hindriks claims, can combine it with this kind of methodological pluralism. In the third chapter, Lauer goes further, arguing that ontological individualism can safely be rejected on the basis of the kinds of inferences we can draw from empirical models in cultural evolution.

These various perspectives, then, point to the need for further work that would explore the limits of methodological individualism and assess the possibility of ontological individualism and holism in cultural evolution. This work can build on the long-standing debates in the philosophy of the social sciences and social ontology on these matters (e.g., Kincaid, 1993; Zahle & Collin, 2014), and would contribute to cultural evolutionary theory by analyzing its assumptions that so far have not been sufficiently explored.

11.2.2 How to Conceptualize Power?

The volume also adds to the literature problematizing power in cultural evolutionary theory and connects to the discussions of power in social ontology and sociology. The contributors offer different views on what constitutes power and how one could study it, thereby pointing to several available venues for further research on the matter.

Baraghith and Jordan (Chapter 6) understand power as a kind of non-imitative social asymmetry which has so far been unaccounted for in cultural evolutionary theory, since the latter is predominantly focused on imitation as a way of perpetuating asymmetries. They propose to integrate this form of power, which they term "structural coercion", into the cultural evolutionary theory by representing it as a kind of niche construction.

Chellappoo (Chapter 5) rejects the above approach as not providing a satisfactory resolution. Her claim is that this approach (as well as other cultural evolutionary strategies of dealing with power: reduction, restriction, and presumption) cannot accommodate structural power relations such as race and gender hierarchies. Such a way of modeling structural inequality, as a network of interactions between individuals, does not offer much insight or explanation of what is actually going on in these cases, Chellappoo claims. In particular, it does not adequately capture the role of higher-level, institutional facts in generating these inequalities, meaning that the scope of these models is limited and may need to be supplemented with explanations in terms of higher-level entities. While Chellappoo presents several cultural evolutionary models of inequality that, according to her, are able to adequately explore some aspects of hierarchical systems, she does concede that these are only partial, how-possibly explanations, not complete functional or causal explanations for the emergence of systems such as racial and gender oppression. Because of this, Chellappoo claims, cultural evolutionary explanations of this kind cannot represent complete accounts of society and culture and replace social scientific approaches.

Burman and Pelgrom (Chapter 7) elaborate on three kinds of power: telic, artisanal, and deontic power. The authors also distinguish between transparent (power that must be represented to exist) and opaque (power that does not need to be represented to exist) forms of social power, claiming that the latter has been overemphasized in both social ontology and cultural evolution. They develop the "opaque forms of social power objection" to cultural evolutionary theory, claiming that the existence of opaque forms of power and their centrality to understanding societies and their evolution present a new challenge to cultural evolutionary theory. According to Burman and Pelgrom, cultural evolutionary theory has difficulty with, or is even unable to, account for this form of social power due to the fact that it is non-reproductive, with agents perhaps being unaware of its existence.

The three differing and partially contradicting views on power put forward in this volume offer possibilities for a further development of a lively discussion of what power in cultural evolution, but also beyond, is and could mean, which forms of power are most central in particular sociocultural domains, and how these are to be represented and accounted for. This discussion also intersects the discussion described earlier on ontological and methodological individualism and holism, with its focus on illuminating individual, institutional, structural, systematic, and other aspects of power.

11.2.3 How to Conceptualize Institutions?

Another prominent issue in the preceding chapters relates to conceptualizing institutions and their evolution. Several partly overlapping and partly competing accounts have been developed in the chapters in the volume.

Hindriks (Chapter 4) offers a rich theory of institutions according to which they consist of equilibria, which are self-enforcing behavioral regularities, and of rules prescribing actions or giving meaning to these actions. Hindriks suggests that cultural evolutionary theory would benefit from incorporating some of the insights provided by this theory of institutions, especially non-value-based attribution of function offered by the notion of rule. Geurts (Chapter 8) presents a rather analogous view, according to which institutions are large, higher-order, often personified correlation devices. In particular, they are clusters of norms for coordinating the actions of their participants, which both create norms and are governed by them. Furthermore, many institutions are agreed to be agents and treated as such by individuals and other institutions.

Stoelhorst (Chapter 10) goes a step further and argues that institutions such as the modern firm are products of gene–culture coevolutionary process and a so-called "major transition" in cultural evolution, in that this form of organization is irreducible to its individual members and their traits. According to this view, this kind of institution is not just a group of people, but an agent in its own right. Roles and norms are what link individual and collective levels of analysis, Stoelhorst claims; they are encoded collective action patterns. With the cultural innovation of legal personhood, we enter a phase of major transition in which cooperative groups are transformed into entities maximizing their own interest, underwritten by our moral psychology, in particular our instinct to behave in line with group norms and roles, which enables these new entities to suppress internal conflict.

It is this psychology and its evolution that Slors focuses on in his chapter (Chapter 9). In particular, he aims to show how self-organizing cycles of socialization can deliver top-down influence in "mindshaping" individuals. In this process, the notion of group-mindedness is crucial in that it presents a combination of naturally evolved sensitivities and culturally evolved practices of socialization, which are crucial to explain human survival in large social groups. Slors also suggests that an analogous process of mindshaping that yields group-mindedness occurs in institutions and underlies their functioning.

These chapters contribute to the theory of institutions in cultural evolution (and beyond), by putting forward accounts of their nature, function, emergence, and persistence. Taken together, they raise questions regarding the most appropriate concept of an institution – as rules-in-equilibria (personified) clusters of norms functioning as correlation devices, or agents in their own right – and related open questions regarding their materialization, operation, and endurance. This marks another area in which cultural evolutionary theory is underdeveloped at present, and where more research is needed in order to bolster the theory and its assumptions.

11.2.4 *Are There Hidden Values, or Alternatively, Blind Spots, in Cultural Evolution, Making It Biased?*

Besides the ontological and epistemological criticisms of cultural evolutionary theory, Lohse and Valković (Chapter 2) also discuss political criticism raised against its various assumptions, tendencies, or implications. In the case of the tendency for individualism, for example, political criticism has been raised against the implication that social inequalities, such as workplace disparities between men and women, are primarily or mostly due to individual choices. When it comes to cultural evolutionary theories' tendency to neglect human agency, arguments have been raised to the effect that this kind of theorizing can negatively affect people who are being represented in such a way by othering or even dehumanizing them. In the case of population-based modeling of organizations and social institutions, political critique points to risks when such thinking is applied in practice, namely that of underestimating the role of intentional actions and encouraging a Social Darwinist view of thriving organizations. Lohse and Valković conclude that evolutionary thinking highlights certain aspects of phenomena while neglecting others, which can often be problematic from political and ethical perspectives. Due to its key theoretical commitments, they claim, a particular reductive worldview enters, if it is not even baked into, cultural evolutionary theorizing, leading to the minimization of the importance of individual human agency when explaining sociocultural phenomena.

The power debate is another area in which the volume raises politically and ethically relevant questions. As mentioned earlier, Burman and Pelgrom (Chapter 7) are strongly critical of cultural evolutionary theory's ability to account for power, in particular, the kind of power they deem central to understanding societies and their development. Baraghith and Jordan (Chapter 6) appear to be on the opposite side of the spectrum, aiming to contribute to cultural evolutionary theory with a novel approach to conceptualizing and modeling power. In her discussion of whether cultural evolutionary theory can accommodate power, Chellappoo (Chapter 5) appears to take a more moderate stance than either of these two contributions, claiming that some individualistic evolutionary game theoretic models can provide valuable insights into structural inequality. She cautions, however, that these accounts should be complemented with explanations in terms of higher-level structures, which could impact their usefulness if, for instance, non-individualistic facts or properties of institutions turn out to be the main drivers of inequality. Differing views on power that these authors present have important implications not only on how power is to be conceptualized within cultural evolution and outside it, but also on the ethical and political considerations in this domain.

The same holds for the views on institutions. Stoelhorst (Chapter 10) argues that institutions such as corporations and states have to be seen as agents that will pursue their own interests, rather than, say, the interests of their shareholders or citizens, or society as a whole. In this way, such an evolutionary ontology of institutions raises obvious ethical and political concerns. The same applies to Slors' (Chapter 9) assertion that perpetuation of a corporation's rules and norms is essentially due to their unreflective automatization on the side of employees. Reflecting on social, political, and ethical aspects and implications of cultural evolutionary theory, including its potential blind spots and biases, presents another possible avenue for research in this domain.

11.3 Evolution and Ontology: Some Connections to Explore at the Intersection of Cultural Evolution and Social Ontology

Further avenues for future research at the intersection of cultural evolution and social ontology can be found when looking more closely at some of the ways in which evolutionary theory is used to mount arguments against or in support of specific views about human behavior and the structures of human societies.[1] In this section, we will discuss the prospects and pitfalls of some of the main reasoning strategies that are used in this context.

11.3.1 Evolutionary History as a Path to Claims About Ontology

An important, albeit indirect, connection between evolution and ontology is historical. Evolutionary biology tells us that the various entities that exist in the living world – not only organisms, but also genes and genomes, cells, colonies, populations, ecosystems, and more – are products of the evolutionary processes that have taken place on Earth during the past 3.5–4 billion years. The strategy followed by many researchers in fields such as evolutionary psychology, evolutionary anthropology, evolutionary sociology, and so on, is to understand the products of evolution by means of elucidating their origins. The basic idea is that it is possible to achieve a better understanding of the nature of organisms through knowledge of the evolutionary trajectories that gave rise to them. This strategy traces back to Darwin's 1871 book, *The Descent of Man*, in which, among other things, he attempted to explain human sociability, cooperativeness, morality, and other traits as products of the evolutionary history of the human species.

While this strategy of reasoning has mainly been applied to study human behavioral and cognitive traits in relation to the biological evolution of the human species, in principle, it can also be applied to study social entities and institutions as products of socio-cultural evolution. While we want to suggest this as yielding possible avenues for future research, we also want to emphasize that these avenues are not without difficulties and are not guaranteed to lead to success. Still, there is a need for research to establish where – if anywhere – they could lead.

A first question is to what extent social entities and institutions are products of intentional human design and to what extent they are products of cultural evolutionary processes. While it is implausible that social entities and institutions can be completely explained as products of cultural evolution (as clearly human design plays *some* role in building such things as governmental institutions, NGOs, economic entities, social structures, and so on), a plausible working assumption is that cultural evolutionary processes play at least *some* role in the origin and subsequent change of social entities and institutions. This is a crucial difference between applying evolutionary theory in the biological domain and applying it in the social domain. In biology, it is traditionally a question of *either* design *or* evolution: the complex functional organization of living beings has long been seen as a phenomenon that requires an explanation invoking an entity who designed living beings, and Darwin presented a naturalistic explanation of organismal complexity that made invoking the actions of a designer obsolete.[2] In the context of cultural evolution and social ontology, however, the situation is different, and a challenge is to distinguish the contribution of cultural evolution to the complex organization of a social entity from the contribution of human design.

A further challenge (in both a negative and a positive sense) pertains to the evolutionary origin of social *kinds*. Darwin's best-known book was about the origin of *species* (but also about the origin of many other biological phenomena – see Reydon, 2025). Biological evolutionary theory encompasses detailed accounts of processes of speciation (e.g., Coyne & Orr, 2004), and much ink has been spilled on the ways in which organisms can and should be grouped into species (for recent assessments of the state of affairs on the "species problem", see Zachos, 2016; Wilkins et al., 2022). This is not the case in cultural evolution, however, and much remains to be clarified regarding the nature of social kinds (what are their membership conditions, are they human or social constructs or closer to natural kinds, and so on – for a detailed discussion, see Epstein, 2025) and, when it comes to cultural evolution, whether it is possible to understand social kinds as products of socio-cultural evolution in a similar way as biological species are understood as products of biological evolution. Our first-order answer to the latter question is negative, as species of organisms and social kinds seem simply too different to be understood in the same way as products of evolution.[3] But this, we believe, should not be taken as a reason to shelve this kind of research before it has even gotten off the ground.

We now turn to two types of evolutionary reasoning that show more concretely how evolutionary history can flow into claims about human behavior and societies, and that are currently widely discussed in the philosophy of science: *evolutionary debunking arguments* (Kahane, 2011; Wilkins & Griffiths, 2013; Machuca, 2023a) and *evolutionary mismatch arguments*. Both types of argument hinge on the claim that aspects of contemporary human life do not fit what humans are as products of billions of years of gradual evolution, but they use this claim in different ways and reach different sorts of conclusions. Both, however, have consequences for views about social ontologies.

11.3.2 Evolutionary Debunking Arguments

Evolutionary debunking arguments are mainly debated in moral philosophy and the philosophy of religion. In ethics, they feature as arguments against moral realism (e.g., Street, 2006; Hopster, 2018; Vavova, 2021), that is, that moral statements express facts and as such can be true or false (for more background on moral realism than we can give here, see, for instance, Shafer-Landau, 2003; Sayre-McCord, 2023). In the philosophy of religion, they feature as arguments against naturalism as well as against the potential truth of religious beliefs (e.g., Plantinga, 1991, 1993; Jong &

Visala, 2014; Thurow, 2023). More recently, evolutionary debunking arguments have also appeared in other areas of philosophy, for instance, in the philosophy of mathematics as arguments against Platonism in mathematics (Leng, 2023), in the philosophy of science as arguments against scientific realism (De Ray, 2022), and more generally, in epistemology as arguments against the truth of commonsense beliefs (Boucher, 2021). We will only briefly discuss evolutionary debunking arguments against moral realism, as that suffices for the purposes of the present chapter.

At first blush, it seems odd to think of moral statements, such as "It is always wrong to lie" or "Thou shalt not steal", as factual statements. What would be the state of affairs "out there" in the world that would make the former statement true in a similar way as the statement "on 1 September 2025 it was a sunny day in East Lansing" is made true by the fact that on that day it was indeed sunny in that particular geographical location? And the latter of the two statements seems even less of a factual claim than the former – it is an imperative or, when taken in a weaker sense, an expression of what a group of people thinks is a wrong kind of behavior. Moral statements do not seem to express states of affairs. But there are good reasons for being a moral realist. Perhaps the most important one is the need for a solid foundation for our moral judgments. We would ideally like moral claims to rest on more than personal opinions and feelings, and on decisions that happen to have been made in the communities in which we live. We would like there to be some objectivity about moral claims – we would like to discover rather than decide that lying is wrong, and thus know that lying is wrong rather than merely agreeing or feeling that it is.

Evolutionary debunking arguments now involve the claim that moral realism is untenable because natural selection only selects for behaviors that promote survival and reproduction, regardless of the veracity of the beliefs that underpin those behaviors. The idea is that if certain behaviors, such as refraining from telling lies or from stealing other people's possessions, happened to help early humans with respect to their survival and reproductive success, then those behaviors as well as the underlying beliefs would be transmitted between generations and become widespread in the population irrespective of any facts of the matter about lying and stealing being wrong. Natural selection only "sees" behaviors, after all, and not the veracity of the underlying beliefs. What matters from an evolutionary perspective is what people *do*. If refraining from stealing promotes survival and reproduction, what matters is that people believe that stealing is wrong and, due to that belief, refrain from stealing. People's beliefs do not have to be true to do their motivating work: as long as people believe that stealing is wrong and act accordingly, their behavior will be selectively advantageous.

While evolutionary debunking arguments feature most prominently in the literature on moral and religious beliefs, it is important to see that they can, in principle, be applied to beliefs in any area. They undermine the epistemic justification of beliefs, that is, they show (if successful) that we do not have sufficient grounds to assume that certain beliefs are true:

> it could be argued that our holding X beliefs can be explained by appealing to the fact that doing so was evolutionarily advantageous for our ancestors and that such an evolutionary advantage did not require our tracking alleged X facts. If so, then, if any X facts do exist, it would be a highly lucky coincidence that our evolutionarily shaped X beliefs matched the X facts. Since we have no epistemic reason to believe that we got lucky, we must suspend judgment about whether our X beliefs are true.
>
> (Machuca, 2023b: 3)

Evolutionary debunking arguments thus lead us from the evolutionary origins of particular human beliefs (or more precisely, the tendency to hold particular beliefs) to the conclusion that these beliefs cannot be assumed to be true (but note that evolutionary debunking arguments do not establish the falsity of beliefs). This general argumentative structure does not depend on *biological* evolution, even though the philosophical literature on the topic usually refers to biological evolution, but can also work on the basis of *cultural* evolution (Machuca, 2023b: 3). This is because the core of this type of argument is that selection processes do not track truth but rather usefulness from the perspective of survival and reproduction.

From the preceding discussion follows a potential connection between cultural evolution and social ontology: can claims about social ontology be evolutionarily debunked? The answer, we believe, is *in principle*, yes. Beliefs about social ontology – and about ontology more generally – can be products of human evolution in the same way as other kinds of beliefs can be. Many beliefs regarding social institutions, social entities, and social structures, such as the institution of marriage and the family, legal systems and the nature of the law, power structures in society, the possibility of attributing responsibility to groups of people as groups rather than only to individual people, the nature of property, and in particular what the various kinds of people are that exist in one's society (Hacking, 2007), directly relate to often deeply rooted ways in which we perceive the world we live in and in which we classify the social world to help us lead our lives. It is thus not implausible to hypothesize that such beliefs have an evolutionary history and that that history might reveal that such beliefs do not stand on solid ground, even though they might be widely held.

But this perspective must be qualified in several respects. Firstly, evolutionary debunking arguments continue to be controversially debated, and it remains unclear how successful they actually can be. The success of an evolutionary debunking argument depends to a considerable extent on the information that is available about the evolutionary history of the human trait (i.e., the tendency to hold a particular belief) under consideration. Often, such information is lacking, and the premises of arguments that rest on the evolutionary origins of a particular trait remain largely speculative (Reydon, 2025: 38–40). It *might* have been the case that the belief that stealing is wrong was selectively advantageous for early humans, but how would we know that this was indeed the case?

Moreover, evolutionary debunking arguments rest on the assumption *that* particular beliefs have evolutionary histories in the first place – an assumption which is difficult to substantiate. A general problem when devising evolutionary explanations of organismal traits is that organisms do not come neatly pre-divided into separate traits that have their own evolutionary history: evolutionary biologists have to individuate traits that they think have a clear evolutionary trajectory and do so on the basis of comparisons between species, that is, by identifying potential homologies that can be traced back to a common ancestor (Reydon, 2023). For beliefs, this clearly is impossible to do such that the assumption *that* a particular belief is mainly due to evolution in early humans must remain speculative, and evolutionary debunking arguments will be hard to get off the ground.

Yet we believe that evolutionary debunking arguments regarding claims in social ontology should not be discarded outright. While for the aforementioned reasons we are skeptical regarding the success potential of evolutionary debunking arguments in social ontology (as well as generally), there is no doubt that human sociality is a product of the biological and cultural evolution of the human species. Therefore, it is at the very least worthwhile to investigate the prospects for evolutionary debunking arguments in social ontology that invoke both biological and cultural evolution. This is an avenue for future research that so far has hardly – if at all – been taken.

11.3.3 Evolutionary Mismatch Arguments

A similar – and perhaps more promising – perspective emerges from considering evolutionary mismatch arguments (Reydon, 2025: 40–41). Evolutionary mismatch arguments have become increasingly popular in psychology, medicine, and public health research as a means to specify a general cause underlying a considerable variety of diseases, disorders, and discomforts, and to provide a basis for interventions to alleviate these (Lea et al., 2023). For example, in evolutionary psychology, authors have invoked evolutionary mismatch arguments to account for chronic psychological stress (Brenner et al., 2015), social media addiction, and other behavioral disorders related to the use of social media and smartphones (Sbarra et al., 2019; Lim & Tan, 2024), disorders and discomforts related to working in largely digital environments (Van Vugt et al., 2024), persisting feelings of dissatisfaction in relation to consumerism (Chang & Durante, 2022), and mental distress in a rural ethnic minority in transition to a modern lifestyle (Speciale, 2021). In public health research, evolutionary mismatch arguments have recently been invoked to explain failures to adequately address the COVID-19 pandemic by reference to inadequate, evolved disease recognition and management capabilities (Seitz et al., 2020; Kessler & Aunger, 2022). In evolutionary medicine, evolutionary mismatch arguments have been advanced to explain various kinds of cancer (Marques et al., 2022; Bieuville et al., 2023), autism, and possibly related immune disorders (Bilbo et al., 2012), differences between the sexes with respect to the occurrence of autoimmune diseases (Keestra et al., 2021), obesity and coronary heart disease (Turaman, 2022), or oral diseases such as caries and gum loss (Smith & Warriner, 2022). These are only a few examples: the bodies of literature in various areas of current psychological, medical, and public health research in which evolutionary mismatch arguments feature, are vast.

Evolutionary mismatch arguments are based on the assumption that human adaptive evolution lags behind environmental change, such that human beings tend to be adapted to the environments of their distant ancestors rather than to the environment in which they currently live. Such adaptation–environment mismatches, the argument goes, are causes for the occurrence of a plethora of diseases, disorders, and discomforts, such as the ones mentioned earlier. Evolutionary mismatch arguments can, in principle, be formulated in terms of mismatches between any ancestral environment and any later environment. Given the focus of psychology, medicine, and public health research on contemporary human conditions, however, the later environment typically is modern industrialized society (Plomp et al., 2022: 8). The ancestral environment is commonly taken to be the Pleistocene and mismatches are often framed as a lack of fit between human traits that have been adapted to living in small hunter-gatherer groups and contemporary life in globalized and heavily industrialized societies. Evolutionary psychologists of what is sometimes called the "Santa Barbara School", for instance, insist that the "Environment of Evolutionary Adaptedness" (Tooby & Cosmides, 1990; see also McCaffree, 2022: 3–24), that is, the environment in which allegedly the evolution of a large part of human behavior and cognition took place, extends over much of the Pleistocene period – an assumption that has been heavily criticized by biologists and philosophers of biology alike (for a brief discussion, see Reydon, 2025:

39–40). As is the case for evolutionary debunking arguments, there are very good reasons to be skeptical regarding the prospects for success of evolutionary mismatch arguments.

This notwithstanding, evolutionary mismatch arguments open up avenues for research in the area of overlap of cultural evolution and social ontology – most importantly, focusing on the question whether it is at all feasible to mount arguments of this kind in this domain. It is not implausible that mismatches occur between partially evolved social institutions and the requirements of present-day societies, as such mismatches are part and parcel of evolution. The notion of evolutionary mismatches is widely acknowledged in evolutionary biology under the header of "adaptive lag" (Bourrat & Griffiths, 2024). On standard evolutionary theory, the adaptation of an evolving population rests on the occurrence of random mutations, subsequent selection, and eventually the fixation of a novel trait in the population, so it is clear that adaptation takes time. Adaptation is conceived of as a gradual reaction of the population to changes in its environment, such that populations must necessarily lag behind environmental changes. Lloyd et al. (2024) describe such lag as a state of disequilibrium in the evolving population, as the population is not in a stable state of being well-matched to its environment, but under external pressure to evolve further. Adaptive lag thus is a key aspect of evolutionary processes, and while it is clear that populations often lag behind their environments, the important question is *how much* they will tend to lag behind.

For an approach to social entities and institutions as partial products of cultural evolution, this gives rise to a number of questions. As discussed at the beginning of this section, a first question to explore is to what extent social entities and social institutions are products of cultural evolution and to what extent they are products of human design. But given that social entities and social institutions are at least partly products of cultural evolution, the velocity with which cultural evolutionary processes take place comes into view as a factor that determines whether any relevant adaptive lag can be expected to be present in social institutions and entities.[4] The problem of the availability of information about the evolutionary history of the trait under consideration, mentioned earlier with respect to evolutionary debunking arguments, also arises here. It will be somewhat less pressing, however, as the cultural evolution of social entities and institutions has taken place over much shorter timescales than the biological evolution of human traits, such that considerably more information is available about possible selective environments.

In any case, the question of adaptive fit between social entities and the societies in which they exist presents itself on the assumption that sociocultural evolution played at least some part in the genesis of social entities. Evolving entities in an important sense have a "life of their own": they gradually change and adapt to changing environmental circumstances by way of an unguided, "blind" process of selection (as well as several other processes) that may involve some degree of lagging behind these environmental changes. If such adaptive lags occur (and this remains an issue in need of further investigation), this will have important consequences for the functioning of social entities and institutions, and hence will be an important topic of investigation in social ontology.

11.4 Methodological Challenges of Interdisciplinary Research

To conclude this chapter, we want to address some methodological issues that arise from the nature of research in an area where two established domains intersect, namely issues connected to interdisciplinarity. As discussed earlier, work in the intersection of cultural evolution and social ontology connects to a number of established fields of research. This means that research at the intersection of cultural evolution and social ontology will typically involve participants from several academic fields – depending on the specific topic, these may include political scientists, political philosophers, sociologists, anthropologists, philosophers of biology, and others. Research endeavors at this intersection will thus have to address challenges that interdisciplinary projects generally face.

When it comes to such general challenges for interdisciplinary work, we do not have a methodology to offer that could resolve these issues. Each interdisciplinary project has its own unique characteristics and its own issues that have to be addressed to get it to work. For entries into the debates on these issues, we refer readers to the rich philosophical and theoretical work on interdisciplinarity (e.g., Klein, 1990; Mäki, 2016; Frodeman et al., 2017; Schmidt, 2022). Here, we want to focus on a specific aspect of interdisciplinarity that surfaces when cultural evolution and social ontology meet: the pluralism that is internal to both of these fields and that causes challenges for attempts to connect work in the respective fields.

An important aspect of interdisciplinary research is the alignment of the content of the involved fields, in particular of concepts, conceptual frameworks, and theories. Participants in a research project need a common ground for exchanging content, that is, a common language to communicate in and a common theoretical framework within which research is done and findings interpreted. One challenge that follows in this respect is due to the fact that neither cultural evolution nor social ontology is a homogeneous, unified field of research.

"Cultural evolution", for example, is not the name of a unified field of research, but rather an umbrella term for a multitude of approaches to the study of human culture from an evolutionary perspective (Mesoudi, 2011: 25ff.; Richerson & Christiansen, 2013; Lewens, 2015; Prentiss, 2021;

McCaffree, 2022). While some of these approaches are close to the theoretical framework of evolutionary biology, others remain at some distance from evolutionary biology and merely use some elements from evolutionary biology or only use evolutionary concepts as analogies. The broad domain of cultural evolution research partly overlaps with many established evolutionary fields, including evolutionary anthropology, evolutionary archaeology, evolutionary economics, evolutionary linguistics, evolutionary sociology, and others, but cannot be thought of as completely subsuming all these fields. Cultural evolution itself is an interdisciplinary domain of research in which researchers from different corners of the domain often work with different theoretical frameworks, different interpretations of central concepts, and sometimes even with different understandings of the nature of evolutionary processes. Even for projects within cultural evolution, a common language and a common theoretical framework can be lacking.

The same holds for the broad field of social ontology. Epstein (2025: 1) defines the domain as concerned with "the study of the nature of the social world", which encompasses asking "What is . . ." – questions about all aspects of the social world: law, money, marriage, language, gender, corporations, power structures, kinds of people, and much more up to and including the nature of societies themselves.[5] As Epstein puts it, "[f]ill in a social object, a social kind, a social category – and it's a topic for social ontology" (Epstein, 2025: 1). Baker remarks that "there has been little consideration of what exactly social ontology is" (Baker, 2019: 2) and defines it as the study of social individuals, social properties, and social kinds, with social properties including "*living in poverty, being a bureaucracy*, and *participating in political elections* that we all pre-theoretically recognize" (Baker, 2019: 12; emphasis in original). Such a broad understanding of the subject matter of social ontology makes close connections to many other fields necessary. While social ontology is conceived of as a subfield of philosophy (Mikkola, 2021: 31), the social entities and classifications that are being studied originate in other fields of science, scholarship, and practice: in sociology, economics, political science, anthropology, history, and so on; in psychology, psychiatry, medicine, public health, and so on (when particular ways of behaving or particular disorders and diseases are used to group people into kinds); and in all the various practical contexts of everyday life. In the same way as philosophers of a special science must be deeply acquainted with the science that they study (and often hold a degree in that science or have at least some academic training in it), social ontologists should be anchored in the specific domain of the social world that they study. Here, too, we see specific challenges relating to interdisciplinarity arise.

Research projects at the intersection of cultural evolution and social ontology are thus faced with a complex of challenges that include finding a common language and theoretical framework within cultural evolution as well as within social ontology, and forging thorough connections to relevant other fields of science and scholarship. These challenges add to the challenges that arise more generally from attempting to connect the two domains. How can these challenges be met?

One suggestion that we want to make is that research projects at the intersection of cultural evolution and social ontology should be conceived of in a thoroughly interdisciplinary and team-based manner. We believe that this kind of research requires a level of knowledge of, or at least familiarity with, too many areas of science and scholarship to be feasible for monodisciplinarily conceived projects. Accordingly, we envisage such research to be done in teams consisting of a diversity of members that, besides researchers in cultural evolution and social ontology, can include (theoretical) evolutionary biologists, philosophers of biology, philosophers of the social sciences, ontologists/metaphysicians, political scientists, sociologists, archaeologists, clinicians, public health researchers, psychologists, and many others. Only teams that encompass members from the fields that are relevant to the project's particular topic will have the required expertise to carry out the research.

Such interdisciplinary projects can be built using available methods and approaches to facilitating the work of interdisciplinary teams. We are unable to provide a comprehensive overview of these in the context of the present chapter, but limit ourselves to mentioning two examples. One example is the Toolbox Dialogue Initiative, based at Michigan State University, that provides resources and organizes tailor-made workshops to facilitate communication in interdisciplinary research teams (Eigenbrode et al., 2007; O'Rourke & Crowley, 2013; O'Rourke et al., 2016; Hubbs et al., 2021).[6] The approach is based on the assumption that many or most difficulties with respect to communication and integration in interdisciplinary teams have a philosophical nature, that is, that they are rooted in – often implicit – differences in concepts and conceptual frameworks used by different team members. These workshops are aimed at making implicit differences in values, approaches, theoretical frameworks, concepts, and so on, explicit, so that they can be discussed among team members and relevant connections can be forged. Another useful example is what has come to be known as the "science of team science", that is, the systematic study of how team-based research works and how it could work better (see Falk-Krzesinski et al., 2011; Ledford, 2015; Cooke & Hilton, 2015). Team science has emerged recently as a focal topic of research in various fields that study how science works and is explicitly aimed at improving collaborative research. It has been characterized as "an amalgam of conceptual and methodologic strategies aimed at understanding and enhancing the outcomes of large-scale collaborative research and training programs" (Stokols et al., 2008: 577) and provides a growing toolkit for enabling team-based research projects.

11.5 Conclusion

As we hope to have shown, the themes explored in this volume demonstrate how cultural evolution and social ontology can impact and enrich each other, and how engaging with "the other side" can be mutually beneficial. Some of the themes that are especially prominent in this respect include ontological and methodological individualism and holism, power, and institutions. Furthermore, in this chapter, we highlighted how these themes and questions raised in the volume are not only of relevance to the study of cultural evolution and social ontology, but also to adjacent and less adjacent fields, including, but not limited to, philosophy of the social sciences, sociology, anthropology, and political philosophy.

In this chapter, we further reflected on possible avenues for future research intersecting cultural evolution and social ontology, which have not been directly addressed by the previous chapters of the volume, and that can be seen when paying close attention to the ways in which evolutionary theory has been used in arguments for or against particular views on human behavior and societies. These strategies include using the evolutionary history of our species as a basis of claims about ontology, as is often used in fields such as evolutionary psychology, evolutionary anthropology, evolutionary sociology, as well as evolutionary debunking and evolutionary mismatch arguments. While not dismissing any of these strategies in principle, we cautioned that they do not only present possibilities for further research, but also pitfalls.

Finally, we reflected on the nature and specific challenges of working in interdisciplinary teams. In this respect, we suggested that the work in research projects intersecting cultural evolution and social ontology should be done in teams and in a thoroughly interdisciplinary manner, including at least researchers from these two disciplines, but ideally also evolutionary biologists, philosophers of biology, philosophers of the social sciences, sociologists, political scientists, and other relevant experts.

We hope that this sketch – as well as the volume as a whole – will inspire the sort of interdisciplinary team-based research that we believe is needed to forge connections between cultural evolution and social ontology.

Notes

1 For a more detailed discussion, see Reydon (2025).
2 William Paley's famous "watch argument" from 1802 (Paley, 2006) posited the complex functional organization of organisms as evidence for the existence of a designer. Darwin provided an alternative, better explanation (Reydon, 2025: 19–20). The debates have not subsided, however, and vestiges of Paley's view linger on with proponents of so-called "Intelligent Design".
3 For one, biological species are widely seen as having the ontological status of (historical) individuals and not classes, whereas a similar view seems implausible for social kinds.

4 It is often assumed that cultural evolution occurs at higher velocities than biological evolution (Perreault, 2012; Birch & Heyes, 2021). While this would make the occurrence of relevant mismatches less likely, as populations respond faster to environmental changes, it should also be kept in mind that societies (i.e., the selective environment) change faster than natural environments.
5 With respect to kinds of people, it is important to notice that we classify people all the time, sometimes without an explicit intention of classifying or even being conscious of the fact that we treat people as people of a specific kind. Epstein (2025) mentions racial groupings as an example, and Mikkola (2021) mentions "landlord" as a social kind. Other kinds of people spring readily to mind (see also Hacking, 2007): laborer, unemployed, homemaker, immigrant, permanent resident, citizen, and many others.
6 See https://tdi.msu.edu.

References

Baker, L.R. (2019): Just what is social ontology?, *Journal of Social Ontology* 5: 1–12.
Bieuville, M., Faugère, D., Galibert, V., Henard, M., Dujon, A.M., Ujvari, B., Pujol, P., Roche, B. & Thomas, F. (2023): Number of lifetime menses increases breast cancer occurrence in postmenopausal women at high familial risk, *Frontiers in Ecology and Evolution* 11: 912083.
Bilbo, S.D., Jones, J.P. & Parker, W. (2012): Is autism a member of a family of diseases resulting from genetic/cultural mismatches? Implications for treatment and prevention, *Autism Research and Treatment* 2012: 910946.
Birch, J. & Heyes, C. (2021): The cultural evolution of cultural evolution, *Philosophical Transactions of the Royal Society B* 376: 20200051.
Boucher, S.C. (2021): Evolutionary debunking arguments, commonsense and scepticism, *Synthese* 198: 11217–11239.
Bourrat, P. & Griffiths, P. (2024): The idea of mismatch in evolutionary medicine, *British Journal for the Philosophy of Science* 75: 921–946.
Brenner, S.L., Jones, J.P., Rutanen-Whaley, R.H., Parker, W., Flinn, M.V. & Muehlenbein, M.P. (2015): Evolutionary mismatch and chronic psychological stress, *Journal of Evolutionary Medicine* 3: 235885.
Chang, Y. & Durante, M. (2022): Why consumers have everything but happiness: An evolutionary mismatch perspective, *Current Opinion in Psychology* 46: 101347.
Cooke, N.J. & Hilton, M.L. (Eds) (2015): *Enhancing the Effectiveness of Team Science*, Washington, DC: National Academies Press.
Coyne, J.A. & Orr, H.A. (2004): *Speciation*, Sunderland: Sinauer.
De Ray, C. (2022): An evolutionary sceptical challenge to scientific realism, *Erkenntnis* 87: 969–989.
Eigenbrode, S.D., O'Rourke, M., Wulfhorst, J.D., Althoff, D.M., Goldberg, C.S., Merrill, K., Morse, W., Nielsen-Pincus, M., Stephens, J., Winowiecki, L. & Bosque-Pérez, N.A. (2007): Employing philosophical dialogue in collaborative science, *BioScience* 57: 55–64.
Epstein, B. (2025): *Social Ontology*, Cambridge: Cambridge University Press.
Falk-Krzesinski, H.J., Contractor, N., Fiore, S.M., Hall, K.L., Kane, C., Keyton, J., Klein, J.T., Spring, B., Stokols, D. & Trochim, W. (2011): Mapping a research agenda for the science of team science, *Research Evaluation* 20: 145–158.

Frodeman, R., Klein, J.T. & Pacheco, R.C.S. (Eds) (2017): *The Oxford Handbook of Interdisciplinarity* (Second Edition), New York: Oxford University Press.
Hacking, I. (2007): Kinds of people: Moving targets, *Proceedings of the British Academy* 151: 285–318.
Hopster, J. (2018): Evolutionary arguments against moral realism: Why the empirical details matter (and which ones do), *Biology & Philosophy* 33: 41.
Hubbs, G., O'Rourke, M. & Orzack, S.H. (Eds) (2021): *The Toolbox Dialogue Initiative: The Power of Cross-Disciplinary Practice*, Boca Raton: CRC Press.
Jong, J. & Visala, A. (2014): Evolutionary debunking arguments against theism, reconsidered, *International Journal for Philosophy of Religion* 76: 243–258.
Kahane, G. (2011): Evolutionary debunking arguments, *Noûs* 45: 103–125.
Keestra, S.M., Male, V. & Salali, G.D. (2021): Out of balance: The role of evolutionary mismatches in the sex disparity in autoimmune disease, *Medical Hypotheses* 151: 110558.
Kessler, S.E. & Aunger, R. (2022): The evolution of the human healthcare system and implications for understanding our responses to COVID-19, *Evolution, Medicine, and Public Health* 10: 87–107.
Kincaid, H. (1993): The empirical nature of the individualism-holism dispute, *Synthese* 97: 229–247.
Klein, J.T. (1990): *Interdisciplinarity: History, Theory, and Practice*, Detroit: Wayne State University Press.
Lea, A.J., Clark, A.G., Dahl, A.W., Devinsky, O., Garcia, A.R., Golden, C.D., Kamau, J., Kraft, T.S., Lim, Y.A.L., Martins, D.J., Mogoi, D., Pajukanta, P., Perry, G.H., Pontzer, H., Trumble, B.C., Urlacher, S.S., Venkataraman, V.V., Wallace, I.J., Gurven, M., Lieberman, D.J. & Ayroles, J.F. (2023): Applying an evolutionary mismatch framework to understand disease susceptibility, *PLoS Biology* 21: e3002311.
Ledford, H. (2015): Team science, *Nature* 525: 308–311.
Leng, M. (2023): The epistemological challenge to robust mathematical Platonism, in: Machuca, D.E. (Ed): *Evolutionary Debunking Arguments: Ethics, Philosophy of Religion, Philosophy of Mathematics, Metaphysics, and Epistemology*, New York & London: Routledge, pp. 207–232.
Lewens, T. (2015): *Cultural Evolution: Conceptual Challenges*, Oxford: Oxford University Press.
Lim, A.J. & Tan, E. (2024): Social media ills and evolutionary mismatches: A conceptual framework, *Evolutionary Psychological Science* 10: 212–235.
Lloyd, E.A., Rao, S., Wilson, D.S. & Sober, E. (2024): Evolutionary mismatch and how to evaluate it: A basic tutorial, *This View of Life*, 22 October, https://zenodo.org/records/13966113.
Machuca, D.E. (2023a): *Evolutionary Debunking Arguments: Ethics, Philosophy of Religion, Philosophy of Mathematics, Metaphysics, and Epistemology*, New York & London: Routledge.
Machuca, D.E. (2023b): Introduction, in: Machuca, D.E. (Ed): *Evolutionary Debunking Arguments: Ethics, Philosophy of Religion, Philosophy of Mathematics, Metaphysics, and Epistemology*, New York & London: Routledge, pp. 1–12.
Mäki, U. (2016): Philosophy of interdisciplinarity: What? Why? How?, *European Journal for Philosophy of Science* 6: 327–342.
Marques, C., Compton, Z. & Boddy, A.M. (2022): Connecting palaeopathology and evolutionary medicine to cancer research: Past and present, in: Plomp, K.A., Roberts, C.A., Elton, S. & Bentley, G.R. (Eds): *Palaeopathology and*

Evolutionary Medicine: An Integrated Approach, Oxford: Oxford University Press, pp. 239–260.
McCaffree, K. (2022): *Cultural Evolution: The Empirical and Theoretical Landscape*, New York & London: Routledge.
Mesoudi, A. (2011): *Cultural Evolution: How Darwinian Theory Can Explain Human Culture and Synthesize the Social Sciences*, Chicago & London: University of Chicago Press.
Mikkola, M. (2021): Social ontology, in: Khoo, J. & Sterken, R.K. (Eds): *The Routledge Handbook of Social and Political Philosophy of Language*, New York & London: Routledge, pp. 31–41.
O'Rourke, M. & Crowley, S.J. (2013): Philosophical intervention and cross-disciplinary science: The story of the Toolbox Project, *Synthese* 190: 1937–1954.
O'Rourke, M., Crowley, S.J. & Gonnerman, C. (2016): On the nature of cross-disciplinary integration: A philosophical framework, *Studies in History and Philosophy of Biological and Biomedical Sciences* 56: 62–70.
Paley, W. (2006): *Natural Theology, or, Evidence of the Existence and Attributes of the Deity, Collected from the Appearances of Nature* (M.D. Eddy & D. Knight, Eds), Oxford: Oxford University Press.
Perreault, C. (2012): The pace of cultural evolution, *PLoS One* 7: e45150.
Plantinga, A. (1991): When faith and reason clash: Evolution and the Bible, *Christian Scholar's Review* 21: 8–33.
Plantinga, A. (1993): *Warrant and Proper Function*, New York & Oxford: Oxford University Press.
Plomp, K.A., Roberts, C.A., Elton, S. & Bentley, G.R. (2022): What's it all about? A legacy for the next generation of scholars in evolutionary medicine and palaeopathology, in: Plomp, K.A., Roberts, C.A., Elton, S. & Bentley, G.R. (Eds): *Palaeopathology and Evolutionary Medicine: An Integrated Approach*, Oxford: Oxford University Press, pp. 1–16.
Prentiss, A.M. (2021): Theoretical plurality, the extended evolutionary synthesis, and archaeology, *Proceedings of the National Academy of Sciences of the United States of America* 118: e2006564118.
Reydon, T.A.C. (2023): The proper role of history in evolutionary explanations, *Noûs* 57: 162–187.
Reydon, T.A.C. (2025): *The Scope of Evolutionary Thinking*, Cambridge: Cambridge University Press.
Richerson, P.J. & Christiansen, M.H. (2013): Introduction, in: Richerson, P.J. & Christiansen, M.H. (Eds): *Cultural Evolution: Society, Technology, Language, and Religion*, Cambridge, MA: MIT Press, pp. 1–21.
Sayre-McCord, G. (2023): Moral realism, in: Zalta, E.N. & Nodelman, U. (Eds): *The Stanford Encyclopedia of Philosophy* (Winter 2023 Edition), https://plato.stanford.edu/archives/win2023/entries/moral-realism/.
Sbarra, D.A., Briskin, J.L. & Slatcher, R.B. (2019): Smartphones and close relationships: The case for an evolutionary mismatch, *Perspectives on Psychological Science* 14: 596–618.
Schmidt, J.C. (2022): *Philosophy of Interdisciplinarity: Studies in Science, Society and Sustainability*, London & New York: Routledge.
Seitz, B.M., Aktipis, A., Buss, D.M., Alcock, J., Bloom, P., Gelfand, M., Harris, S., Lieberman, D., Horowitz, B.N., Pinker, S., Wilson, D.S. & Haselton, M.G. (2020): The pandemic exposes human nature: 10 evolutionary insights, *Proceedings of the National Academy of Sciences* 117: 27767–27776.
Shafer-Landau, R. (2003): *Moral Realism: A Defense*, Oxford: Oxford University Press.

Smith, T.M. & Warriner, C. (2022): Developmental, evolutionary and behavioural perspectives on oral health, in: Plomp, K.A., Roberts, C.A., Elton, S. & Bentley, G.R. (Eds): *Palaeopathology and Evolutionary Medicine: An Integrated Approach*, Oxford: Oxford University Press, pp. 95–119.
Speciale, A.C. (2021): Prevalence and factors related to psychological distress among ethnic minority adults in a semi-modern village in rural Vietnam: An evolutionary mismatch framework, *Evolution, Medicine, and Public Health* 9: 194–205.
Stokols, D., Hall, K.L., Taylor, B.K. & Moser, R.P. (2008): The science of team science: Overview of the field and introduction to the supplement, *American Journal of Preventive Medicine* 35: S77–S89.
Street, S. (2006): A Darwinian dilemma for realist theories of value, *Philosophical Studies* 127: 109–166.
Thurow, J.C. (2023): Debunking arguments and religious belief, in: Fuqua, J., Greco, J. & McNabb, T.D. (Eds): *The Cambridge Handbook of Religious Epistemology*, Cambridge: Cambridge University Press, pp. 290–304.
Tooby, L. & Cosmides, L. (1990): The past explains the present: Emotional adaptations and the structure of ancestral environments, *Ethology and Sociobiology* 11: 375–424.
Turaman, C. (2022): Classification of the risk factors of coronary heart disease and their evolutionary origins, *Health Sciences Review* 3: 100027.
Van Vugt, M., Colarelli, S.M. & Li, N.P. (2024): Digitally connected, evolutionarily wired: An evolutionary mismatch perspective on digital work, *Organizational Psychology Review* 14: 403–424.
Vavova, K. (2021): The limits of rational belief revision: A dilemma for the Darwinian debunker, *Noûs* 55: 717–734.
Wilkins, J.S. & Griffiths, P.E. (2013): Evolutionary debunking arguments in three domains: Fact, value, and religion, in: Dawes, G.W. & Maclaurin, J. (Eds): *A New Science of Religion*, New York & London: Routledge, pp. 133–146.
Wilkins, J.S., Zachos, F.E. & Pavlinov, I.Y. (Eds) (2022): *Species Problems and beyond: Contemporary Issues in Philosophy and Practice*, Boca Raton: CRC Press.
Zachos, F.E. (2016): *Species Concepts in Biology: Historical Development, Theoretical Foundations and Practical Relevance*, Cham: Springer.
Zahle, J. & Collin, F. (Eds) (2014): *Rethinking the Individualism-Holism Debate: Essays in the Philosophy of Social Science*, Cham: Springer.

Index

Note: Page numbers in *italics* indicate a figure and page numbers in **bold** indicate a table on the corresponding page.

For Product Safety Concerns and Information please contact our EU
representative GPSR@taylorandfrancis.com
Taylor & Francis Verlag GmbH, Kaufingerstraße 24, 80331 München, Germany

www.ingramcontent.com/pod-product-compliance
Ingram Content Group UK Ltd.
Pitfield, Milton Keynes, MK11 3LW, UK
UKHW022312250926
475702UK00046B/733

* 9 7 8 1 0 4 1 0 8 1 2 1 0 *